De-Extinction and the Genomics Revolution

Amy Lynn Fletcher

De-Extinction and the Genomics Revolution

Life on Demand

Amy Lynn Fletcher
Christchurch, New Zealand

ISBN 978-3-030-25788-0   ISBN 978-3-030-25789-7   (eBook)
https://doi.org/10.1007/978-3-030-25789-7

Cover illustration: Pattern © Melisa Hasan

This Palgrave Pivot imprint is published by the registered company Springer Nature Switzerland AG.
The registered company address is: Gewerbestrasse 11, 6330 Cham, Switzerland

This book is dedicated to my dad, William Miller Fletcher, Jr. (1938–2018). Every word in this that is mine also belongs to you.

CONTENTS

LIST OF IMAGES

CHAPTER 1

Preface and Acknowledgments

Abstract This short book considers the implications of "life-on-demand," that is, our ability to manipulate life via biotechnology. The enquiry centers on the idea of de-extinction, which gained global visibility in 2013 due to a TEDx De-Extinction event sponsored by several deep-pocketed stakeholders, but has its genesis in the scientific and social fascination with life, death, and immortality that dates to the nineteenth century and has persisted in various guises ever since. Against the backdrop of the contemporary extinction crisis, leveraging the power of biotechnology is arguably crucial to sustainability and even to human survival. Yet the dream of herds of mammoths trundling across the tundra or flocks of passenger pigeons darkening the skies again pulls on the imagination in ways that go far beyond science. This book thus scrutinizes the porous boundaries between the possible and the probable, fact and fantasy, inevitability and hype, that constitute science in society in the twenty-first century.

Keywords Biotechnology • De-extinction • Immortality • Silicon Valley • Science in society • Socio-technical Imaginaries

This book considers the implications of nature made to order or what I call life-on-demand. As digital and biotechnologies extend the frontiers of what is possible with respect to human manipulation of the natural environment and species, including our own, it is important to pause and consider both

© The Author(s) 2020
A. L. Fletcher, *De-Extinction and the Genomics Revolution*,
https://doi.org/10.1007/978-3-030-25789-7_1

1

how we reached this moment and what it signifies about how human beings relate to both the contemporary extinction crisis and the environmental imaginary of nature yet to come. The enquiry centers on the idea of de-extinction, which gained global visibility in 2013 due to a major TEDx De-Extinction event sponsored by several deep-pocketed stakeholders, but has its genesis in the scientific and social fascination with life, death, and immortality that gripped the nineteenth century in Europe and the United States and which has persisted in various forms ever since.

De-extinction, and its more established conceptual cousin, genetic rescue, are two ideas that deserve to be taken seriously as innovative responses to the catastrophic extinction crisis that surrounds us. Genetic rescue seeks to maximize genetic diversity and minimize inbreeding through either natural or facilitated reproduction in small and isolated populations. Genetic rescue "is a tool that can stem biodiversity loss more than has been appreciated, provides population resilience, and will become increasingly useful if integrated with molecular advances in population genomics."[1] Against the stark reality of the contemporary extinction crisis, wherein approximately 27 percent of the world's species are threatened,[2] genetic rescue could improve conservation interventions, fortify population resilience, and stem biodiversity loss.[3] In the mid-1980s, Michael Soulé described conservation biology as a crisis discipline, arguing that conservation biology's relationship to biology is analogous to "that of surgery to physiology and war to political science."[4] In the decades since, the extinction crisis has only worsened and is now reaching potentially catastrophic tipping points. Leveraging the power of genetic rescue and population genomics is crucial to averting this ecological disaster.

Yet as genetic rescue integrates within mainstream conservation work, a complementary idea, de-extinction, generates a steady stream of controversy, mass media attention, and public fascination since, at the farthest end of their scientific and entrepreneurial ambition, de-extinction proponents raise the possibility that advanced biotechnological tools such as reproductive cloning, gene editing, and ancient DNA analysis provide a means not only to revitalize extant populations but to bring back extinct species such as the woolly mammoth, Tasmanian tiger, or passenger pigeon. Critics of de-extinction worry about the diversion of scarce resources to such speculative "resurrection" projects and argue that the preservation of endangered species should be prioritized over fanciful projects to bring back charismatic lost species that departed the Earth quite some time ago. Proponents counter that de-extinction will actually contribute to deep ecological enrichment of the planet by restoring key-stone species that humans carelessly decimated in earlier eras and that the long-term research will yield insights that may also save critically endangered species from passing over the threshold into extinction. As the field of de-extinction takes shape, other issues that have been well canvassed over the last decade include moral hazard, wherein the notion that species can be readily "brought back" reduces the incentive to protect them in the first place, the question of species authenticity, given that any "resurrected" species would technically be a hybrid (due to the use of surrogate species as either a genome template or for gestation), and the bioethical implications of bringing back a few remnants of species (such as woolly mammoths) that in their own time traveled in large herds and had fine-tuned social behaviors.

This book shifts focus from the arguments for and against de-extinction to consider its broader promissory and cultural dimensions. First, as will be seen, the idea of de-extinction is not quite as new as some of its contemporary proponents profess it to be, having a pedigree that tracks back to the emergence of the science fiction novel in the nineteenth century and to several influential scientific figures in the history of modern biology. Intimations of creating life-on-demand can be found in the nineteenth-century musings of physicist Ernst Mach (1838–1916) and biologist Jacques Loeb (1859–1924), for example, and, in 1951, the German scientist Heinz Heck (1894–1982) published accounts of his attempts to breed back the auroch—the progenitor of most contemporary cattle that, as a distinct species, had died out by 1667. This book places the contemporary subject of de-extinction and the media discourse scaffolding it into this broader historical context. It demonstrates how the twenty-first-century environmental imaginary, in which life is made to order through mastery of technology and seemingly fixed borders such as the line between extant and extinct become much more porous and open to human intervention, is firmly rooted in aspirations to exert control over nature that characterize the modern biotechnological project.

In doing this, I draw upon summative historical evidence as well as the vast scientific and popular literature on de-extinction that has emerged since the mid-1980s, when Russ Higuchi and his research team at the University of California released the first evidence that so-called ancient DNA could persist, in fragmented form, for at least 150 years.[5] Analyzing the voluminous amount of documents, interviews, articles, and presentations that underwrite de-extinction discourse allows one to evaluate how the prospect of "resurrection biology" emerged, how it has been critiqued and negotiated in the mass media, and what visions of the future of wilderness and our technological relationship to nature are in play. As Alan Petersen argues, "because the media operate at the interface between genetic researchers and the public, they are likely to play an important role in shaping public perceptions of genetics and its value and applications, by selectively presenting some subthemes and not others."[6]

The idea of de-extinction emerges from serious science but it also captures the imagination of the mass media, the general public, novelists, and film-makers in multiple ways that ensure that the boundary between science fact and science fiction remains permeable. Ergo, I also consider how disputants in this debate negotiate the limits of de-extinction in the public domain (inclusive of digital and print media, TED conferences, the published scientific literature, and other outlets for scientific research and debate) and navigate the tension between the desire to normalize de-extinction within conservation science and the deployment of resurrected woolly mammoths to attract support, attention, and funding within a crowded scientific and media landscape. The organizing premise of this book is that de-extinction is a rich case study of the tension between regimes of hope, in which new and better scientific interventions and products are always on the horizon, and regimes of truth, in which science must recalibrate the timeline for success and invoke expertise to demarcate the line between science and hype.[7] Just as stem cell research gained widespread public attention and funding traction in the early 2000s, when put in the promissory context of curing degenerative diseases, the image of herds of mammoths trundling across the tundra or flocks of passenger pigeons darkening the skies again pulls on the human imagination and bridges the knowledge gaps between scientists, investors, and the mass public. This fact does not imply deliberate manipulation by any stakeholder in this debate. Instead, an essential starting point in this analysis is that the contemporary ecosystem of big science, inclusive of multiple and interlocking actors and agendas and sustained by the public's attention and support, inevitably blurs the line between science and science fiction. We therefore have to parse continuously the shifting boundaries between the possible and the probable, fact and fantasy, inevitability and hype.

Multiple players engage daily in constructing the field of de-extinction— entrepreneurs, scientists, ethicists, journalists, artists, academics, and investors—and collectively the interaction between these stakeholders advances a new discourse of life and conservation that renders old edicts such as "when a species is gone, it is gone forever" less absolute. Moreover, in an era of exponential technological change, wherein the power to shape politics, culture, and markets shifts to prominent technology centers such as Silicon Valley, certain corporate players can push the frontiers of the conceivable more consequentially than others. The debate about the benefits, risks, and ethics of de-extinction thus intersects with the imperative to extract applied and commercial value from research in order to keep the biotechnological innovation ecosystem alive and functioning. As the story is told, a few enterprising and well-connected individuals had a disruptive idea: what if extinction need not remain the final stop on the evolutionary journey? Through a combination of grit, enthusiasm, talent, and research, stakeholders invested in this idea decided to find out if we could eventually replenish the Earth with vanished species. As with all such stories, this account is partially true, but it obscures the more tangled pathway via which certain elite ideas and agendas gain traction in science and society and others do not. Ergo, tracing the idea of de-extinction as it shifts between the modes of normal science and spectacular science, disruptive innovation and incremental innovation, and aspirations and retrenchment is to elucidate a paradigmatic case of life-in-the-making.

This analysis attends to the discursive politics of de-extinction, the way in which texts, documents, speeches, and scientific talks combine to produce a new framework both for wilderness restoration and for thinking about our relationship to the natural world. Per John Dryzek, "a discourse is a shared way of apprehending the world. Embedded in language…each discourse rests on assumptions, judgements, and contentions that provide the basic terms for analysis, debates, agreements, and disagreements, in the environmental area."[8] A focus on how language is used to create new scientific possibilities, and on the friction between public language and expert language, demonstrates how de-extinction is one key territory in a much larger biotechnological frontier being explored in this century. In applying the tools of narrative analysis to this case, I accept the position that "language—whether it emanates from scientific reports, congressional testimonies, or Internet sites—is a strategically crafted story that frames issues in ways that produce desired outcomes."[9] Moreover, as the contemporary debate about de-extinction parallels the rapid proliferation of social media, alternative media sites can be added to the category of public consumption documents. Discursive analysis holds that ideas and language matter. In the case of de-extinction, competing ideas about the causes and consequences of the present environmental crisis, the human role in accelerating this crisis, and the proper use of biotechnology to stem the crisis and possibly reverse it interact with the slow accumulation of empirical facts. De-extinction, as an applied science, also interacts with contemporary frameworks such as eco-modernism, rewilding, and the Anthropocene, all of which seek, in different ways, to reconcile the ethical, technological, and political demands of the present environmental moment.

Finally, this book is indebted to the rich literature on socio-technical imaginaries. While de-extinction is not a national socio-technical imaginary in the strictest sense, such as Cold War spaceflight in the 1960s or nuclear power in France in the 1970s, it is an imaginary coming into view around life, death, and, at the farthest end of possibility, immortality. To garner our support for de-extinction in the present, advocates mobilize the environmental future and draw a promissory future nature filled with majestic creatures from the past such as woolly mammoths, giant moas, and saber-tooth tigers. In this sense, de-extinction becomes one of the new technologies of life whose proponents leverage "their forward vision: they seek to optimize the vital future by action in the vital present."[10] This is why, as will be discussed, much of the scientific frustration with social responses to de-extinction misses the mark. There is, of course, a set of facts against which de-extinction must be measured in the world of elite science, but when one party to the public discussion begins by promising to bring back woolly mammoths, it becomes difficult to then pull up the drawbridge and retreat behind walls of pure science after the audience counters with ideas and concerns of its own.

To elaborate upon these ideas, the book is organized as follows. Chapter 2 begins with the story of Sudan, the last known male northern white rhino in existence, who finally departed the Earth in 2018. From this foundation, it moves back in time to contextualize de-extinction as an idea that has its essential origins in the early nineteenth century, which witnessed the rise of the vibrant new sciences of evolution and biology and the parallel emergence of the science fiction novel and popular scientific spectacles. Chapter 3 continues this cultural and historical contextualization of de-extinction by evaluating the rise of contemporary molecular biology, dating approximately to Watson and Crick's discovery of the double-helical structure of DNA, and the powerful influence of the life-as-code metaphor on the possibilities envisioned for biotechnology. Chapter 4 considers the woolly mammoth as a key de-extinction media icon that readily shifts temporally between the Pleistocene and the Anthropocene and wraps a beguiling image and intriguing narrative around otherwise abstract and often difficult to comprehend scientific concepts. Chapter 5 connects the promissory notion of herds of reanimated woolly mammoths to the centrality of synthetic biology and gene editing in the twenty-first century. It considers the ghosts of mammoths yet to come and the way in which de-extinction as an idea and a promise connects to the ongoing search in some unorthodox scientific circles even for the elixir of immortality. When life is made to order, no boundaries seem altogether impermeable (Image 1.1).

Image 1.1   The age of monsters. Frontispiece, *The Fairy Tales of Science*, 1859, by John Cargill Brough (Illustration by Charles H. Bennett)

In 1859, a Victorian gentleman named John Cargill Brough published a volume entitled *The Fairy Tales of Science: The Age of Monsters*, described as a book for youth in which "the revelations of modern science transcend the wildest dreams of the old poets; and in exchange for a few shadowy griffins and dragons, we are presented with a whole host of monsters, real and tangible monsters too, who in the early days of the world's history were the monarchs of all they surveyed."[11] Mr. Brough seemed to be picking up, with rather a lot of literary license, on many of the same scientific reverberations that would soon find their consolidation in Charles Darwin's epochal *On the Origin of Species by Means of Natural Selection*. Of course, Darwin's book ranks among the most significant scientific accomplishments of his or any age, while *The Fairy Tales of Science* seems fated to rest quietly in the digital archives of Wikisource and perhaps in a few libraries scattered around the world. Still, Mr. Brough did manage to compose perhaps the most sublime and succinct acknowledgments in the English tradition, and it is in his honor as an early exemplar of the popular study of science that this opening chapter concludes: "that I have been obliged, in the composition of the work, to consult a crowd of authorities, need hardly be stated, nor will any more formal enumeration or systematic acknowledgment be expected."[12]

Notes

1. Whiteley, Fitzpatrick, Funk, Tallmon, "Genetic Rescue to the Rescue," 42.
2. International Union for the Conservation of Nature, 2018.
3. Whiteley, et al., 42.
4. Soulé, "What Is Conservation Biology," 727.
5. Higuchi, Bowman, Frieburger, Ryder, Wilson, "DNA Sequences from the Quagga, an Extinct Member of the Horse Family."
6. Petersen, "Biofantasies: Genetics and Medicine in the Print News Media," 1256.
7. Moreira and Pallidino, "Between Truth and Hope," 67.
8. Dryzek, *Politics of the Earth: Environmental Discourses*, 8.
9. McBeth, Shanahan, Jones, "The Science of Storytelling: Measuring Policy Beliefs in Greater Yellowstone," 414.
10. Rose, *The Politics of Life Itself: Biomedicine, Power, and Subjectivity in the Twenty-First Century*, 8.
11. https://en.wikisource.org/wiki/The_fairy_tales_of_science/The_Age_of_Monsters.
12. Ibid.

CHAPTER 2

Matters of Life and Death
in the Anthropocene

Abstract This chapter begins with Sudan, the last male northern white rhino, in order to put a face on both the extinction crisis and the conservation potential of de-extinction. It then shifts to the eighteenth century wherein our desire to transcend the boundary between life and death flourished, and professional science and popular science began the slow, uneven process of separating into distinct domains. The chapter concludes that de-extinction extends this earlier tradition, while becoming the quintessential environmental imaginary of our biocybernetic age, one in which the key lesson of *Jurassic Park* may lie not in scientific hubris or creatures run amok but rather in demonstrating that science and science fiction invariably double back on each other in the desire to restore lost worlds.

Keywords Anthropocene • De-extinction • Biocybernetics
• Extinction crisis • *Jurassic Park* • Northern White Rhino

Sudan, the last known male northern white rhino (*Ceratotherium simum ssp. cottoni*) on earth, died on March 30, 2018, at the Ol Pejeta Conservancy in Kenya. Though the southern white rhino (*Ceratotherium simum ssp. simum*) was nearly exterminated on the continent of Africa in the colonial era, twentieth-century wildlife conservation programs, which worked from a small extant population of less than 100 animals found in 1895 in Kwazulu-Natal, South Africa, stabilized the population to approximately

© The Author(s) 2020
A. L. Fletcher, *De-Extinction and the Genomics Revolution*,
https://doi.org/10.1007/978-3-030-25789-7_2

20,000 extant rhinos today. In contrast, while in 1960 an estimated 2000 northern white rhinos could be found ranging across a territory that included southern Chad, the Central African Republic, southwestern Sudan, the Democratic Republic of the Congo, and Uganda, after 1970 the numbers fell precipitously and never had an opportunity to recover.[1] With prices for illegal rhino horn reaching an estimated $100,000 USD per kilo in the early 2000s, an escalation driven primarily by the demand for rhino horn as an ingredient in traditional Asian medicine in countries such as China, poaching hit the northern white rhino population, already acutely vulnerable at this point, particularly hard, such that extinction became essentially inevitable.[2] As Dr. Oliver Ryder, the Kleberg Endowed Director of Conservation Genetics at the San Diego Zoo notes, dwindling populations can reach a point of extinction inevitability at which "the numbers get smaller and it becomes a feedback loop. Like water going down a drain: It's deterministic. Or a mass entering a black hole. Once you hit the event horizon, you're out of here."[3] Sudan's death left only two female northern white rhinos behind, named Fatu and Najin, who live under constant surveillance at the Conservancy but are unable to breed naturally with males of the related subspecies. *C. simum cottoni* is currently listed as functionally extinct by the International Union for the Conservation of Nature (IUCN).

Yet hope for the species may lie in the frozen tissue samples held in the San Diego Frozen Zoo, which could potentially lead to the use of stem cell technology to create eggs and sperm from skin tissue and then eventual implantation of a northern white embryo into a southern white rhino surrogate. Genetic rescue of the northern white rhino, while scientifically uncertain and expensive, could be, according to conservation scientist Joseph Bennett, an inspirational "good news story for people"[4] and for the species. In July 2018, news reports around the globe reported that AVANTEA, an assisted reproductive technology company based in Italy (in collaboration with Dvůr Králové Zoo, the Leibniz Zoo, and the Wildlife Research Institute [IZW] in Berlin), had created a viable hybrid in vitro embryo by combining a single northern white rhino sperm with an egg harvested from a southern white rhino. A research team at Scripps Research Institute, led by scientist Jeanne Loring, is pursuing an alternative research program that hopes to use induced pluripotent stem cells (iPS) technology for assisted reproduction efforts to reintroduce sufficient genetic diversity into a revived northern white rhino population.[5]

Not everyone, however, is convinced that these biotechnological interventions will prove to be able to restart a viable northern white rhino population, nor that a biotechnological approach can redress the fundamental issues that caused the northern white rhino to disappear in the first place. Save the Rhino, an environmental non-governmental organization (NGO), emphasizes that even were the subspecies to be revitalized via assisted reproduction (which is unlikely in the near-to-medium term, given that only ten in vitro fertilisation (IVF) rhino births took place between 2002 and 2017), there is still no place for them to roam safely and concludes that "the best outcome will be to put our efforts and funding—including research into IVF—into saving the species which do still have a chance," such as the critically endangered Javan (*Rhinoceros sondaicus*) and Sumatran (*Dicerorhinus sumatrensis*) rhinos.[6] Moreover, as conservationist Stuart Pimm reminds us, "we still live in a world in which we have lost an enormous number of rhinos to poaching, and if we have any chance of putting their descendants back into the wild, we'll have to prevent them from being killed the moment they're released."[7]

This debate about the relative merits of high-tech biotechnological interventions versus the seemingly intractable political and economic challenges of animal poaching and loss of habitat encapsulates the brutal dilemma facing many species today. The fate of the northern white rhino is particularly dire, but it reflects the severity of an escalating extinction crisis occurring around the globe. Urbanization, intensive agriculture, deforestation, and anthropogenic climate change are combining to produce a sixth mass extinction event—one in which human actions and desires play a leading role in the destruction and accelerate the extinction clock far beyond the normal baseline extinction rate of natural evolution. A widely publicized analysis of entomofauna, for example, found that 40 percent of the world's insect species are threatened with extinction, including butterflies, moths, and honeybees.[8] The Shark Specialist Group at the IUCN classifies 17 shark species as being either vulnerable, endangered, or critically endangered, largely as a result of humans overfishing the oceans.[9] Over the last 50 years, an estimated 90 amphibian extinctions have occurred, while chytridiomycosis, a fungal disease, further decimates global amphibian populations.[10] The 2016 IUCN Red List identified 205 critically endangered mammalian species, including the Eastern black rhinoceros (*Diceros bicornis michaeli*), Sumatran orangutan (*Pongo abelii*), Western lowland gorilla (*Gorilla gorilla ssp. gorilla*), 22 different species of lemur, and the Sumatran tiger (*Panthera tigris sumatrae*).

Economic pressures to industrialize and the seemingly inexorable expansion of human communities into the last remaining preserves of wild animals raise the real prospect of a world that endures without any megafauna, and without most of the world's amphibians, insects, or reptiles, by the twenty-first century.

In light of this biodiversity catastrophe, it only makes sense that multiple stakeholders would look to advanced digital and biotechnologies and the promissory suite of potential tools each may offer as a way out of this mess, and that some can even envision a world not only saved from the brink but replenished with revivified extinct species. As with the various spin-off research projects enabled by the Human Genome Project (1990–2003) in the complementary realm of human health, there seems to be a treasure trove of scientific knowledge now contained in laboratories and research papers that is just on the verge of translating into real-world applications that could halt the extinction crisis, if only we could find the key to unlock it. Even de-extinction critics, such as University of Otago zoologist Phil Seddon, acknowledge that de-extinction is:

> not so much a single technical advance (though the new gene-editing tool CRISPR will be transformative), as a coming together of developing techniques that make a new application possible—and it is possible, make no mistake. The next decade will see the cloning or genetic reconstruction of some version of a formerly extinct species; one that will live long enough to breathe and shake its fur, feathers, or scales, or to unfurl a leaf. The technical challenges that remain are formidable, but so very much has already been achieved along the de-extinction pathway.[11]

This chapter tracks the contemporary de-extinction pathway back to its roots in the late eighteenth century and the early stages of what became modern science and technology. Though the specific idea of de-extinction is disruptive, in that it unsettles established conservation paradigms and introduces new players into the professional space of environmental preservation, the desire to exert human control over the processes of life and death—and to hold onto a vital past—is a preoccupation that weaves through Western culture and society since at least the Enlightenment era. Indeed, early notions of the atom as the essential building block of matter can be credited, broadly, to Democritus (c. 470–c. 360 BC), whose insights, while ridiculed by Aristotle and ignored for approximately 2000 years, anticipate the influential work of the English scientist John Dalton on atomic theory in chemistry (1766–1844). Indeed, it is at the juncture between the eighteenth and nineteenth centuries, as Georges Cuvier (1769–1832) establishes the scientific fact of extinction that questions about life, death, the nature of matter, and evolution begin to roil scientists, intellectuals, and artists. It is here, as well, that the essential foundation for the twenty-first-century idea of de-extinction is poured and the Janus-faced themes of science as redemption and science as vanity catapult into broad public awareness through the efforts of learned societies, amateur scientists, early mass media, novels, and theatrical displays. The chapter concludes by returning to the nascent twenty-first century in order to consider core ideas such as rewilding and the Anthropocene and to demonstrate how these ideas, particularly as they intersect with popular tropes such as *Jurassic Park*, underwrite de-extinction as the quintessential environmental imaginary of our biocybernetic age, one in which science and science fiction repeatedly double back on each other in the urge to restore lost worlds.

LIFE, DEATH, AND LIFE (AGAIN) IN THE MODERN IMAGINATION

Michael Crichton's novel *Jurassic Park* (1993) concerns an enterprising businessman who launches an isolated island theme park replete with cloned dinosaurs whose number includes at least one enormous and hungry *Tyrannosaurus rex*. The novel provided not only the source material for one of the most successful movie franchises of all time but also a widely recognized popular metaphor for journalists seeking to convey the new science of de-extinction to contemporary mass audiences. Throughout hundreds of mass media articles reporting on various de-extinction breakthroughs and initiatives in the period 1999–2018, references to *Jurassic Park* far outnumber any other trope. In Crichton's imaginary landscape, dinosaurs are brought back to life via the extraction of dinosaur DNA (deoxyribonucleic acid) from specimens that were fossilized in amber that is then interjected into frog embryos in the laboratory. In what one writer refers to as "Jurassic Park's unlikely symbiosis with real-world science,"[12] just one day prior to the launch of the movie, the esteemed science journal *Nature* published a scientific study which reported the successful extraction of DNA from a weevil estimated to be approximately 130 million years old that had been resting through the millennia encased in Lebanese amber.[13] Though subsequent studies, such as one published in *Molecular Biology and Evolution* in 1998, cast doubt on the findings and hypothesized that the material found in the amber was actually fungal contamination,[14] the tantalizing connection between the normal science of ancient DNA and the new possibilities of the emerging biotech era were forged in the public's mind and helped to propel the original movie and the subsequent franchise (which includes not only sequels but also theme parks and memorabilia) into the realm of multi-billion-dollar worldwide gross profits. In the 1993 original, the smug lead scientist, Dr. Henry Wu, reassures the skeptical scientists who have been assembled to critique the project that only females are being brought back, so there is absolutely no chance that the animals will begin to reproduce naturally and kick-start an uncontrolled dinosaur population some 65 million years out of its own time. The catch, of course, as the bemused chaos theoretician reminds us (taking a bit of scientific license), is that frogs are hermaphroditic and can, therefore, take either the male or female sex role in reproduction, ergo … one need only wait for the unforeseen but inevitable disaster to ensue.

Jurassic Park (1993) captured the anxieties and anticipations of a society in which major initiatives such as the Human Genome Project were focusing mass attention on the promise of biotechnological research but also on the possibility of unethical uses such as attempted human cloning. As biotechnological progress has continued apace, the ongoing popularity of the movie franchise, wherein each installment relies for dramatic effect on dinosaurs of escalating size and ferocity, and various villains seek either to commercialize or weaponize these benighted creatures, suggests that the essential storyline taps into something primal in us, can momentarily transport us back to the campfires where we first heard these scary monster stories that now hover in the back of our collective consciousness. While Crichton may not have gotten every scientific detail right, the story is a brilliant mixture of the serious and the populist, and itself resurrects themes inherited from sources that range from Arthur Conan Doyle's *The Lost World* (1912) to sci-fi drive-in movies of the mid-twentieth century such as the delightfully execrable *King Dinosaur* (1955), in which scientists leave Earth to explore a distant planet and decide to nuke the dinosaurs that live there in order to make room for eventual human habitation. Certainly, in its focus on delusional (if not mad) scientists and on dangerous reanimated creatures run amok, Crichton's work also respectfully tips the hat to Mary

Shelley's formative *Frankenstein, or the Modern Prometheus* (1818), in which scientist Victor Frankenstein tragically brings a monster to life through a combination of chemistry, alchemy, and an electrical jolt. Her novel launches the modern genre of science fiction and sketches an endlessly compelling portrait of the tension between scientific curiosity and hubris, between knowledge and disaster, between life and death. Just as Crichton's novel seems now to be the inevitable popular corollary to the emergent Biotechnological Age, Shelley's novel emerged from within a vibrant intellectual climate in which questions about the composition of matter, the uses of galvanism, and the question of where to locate the boundary between life and death were all subject to intense debate and scientific investigation.

The early nineteenth century is also an era wherein the boundaries erected in the twentieth century between professional science and popular culture were much weaker, indeed only in the nascent stages of being established at all. As Bernard Lightman argues, "the distinction between the popularizer of science and the science practitioner should not be seen as being too rigid" prior to the twentieth century.[15] In that context, policing the veracity of *Frankenstein* arguably counted for less than trying to assimilate the larger points it was making about human intervention in natural processes and the aspirations of modern science. The surgeon, John Abernathy, President of the Royal College of Surgeons, for example, contended in a famous 1814 debate with William Lawrence on the nature of life, that "life did not depend upon the body's structure, the way it was organized or arranged, but existed separately as a material substance, a kind of vital principle, 'superadded' to the body."[16] This notion of a vital principle, though here spiritually grounded and not moored in biological science, faintly echoes the scientific preoccupation with a question that would command the attention of physicist Erwin Schrödinger in the 1940s: what is life? It also had a counterpart in the nineteenth-century study of galvanism—chemically generated electricity—that likewise found its way into Mary Shelley's dreamscape. The physicist Giovanni Aldini (1762–1834), in an effort to confirm and publicize his uncle Luigi Galvani's (1737–1798) theory of "animal electricity," by which he meant an innate vital force inherent in living beings, conducted widely publicized popular demonstrations of electrical jolts being applied to piles of oxen or dog specimens, or even to the cadavers of recently executed criminals (Image 2.1).

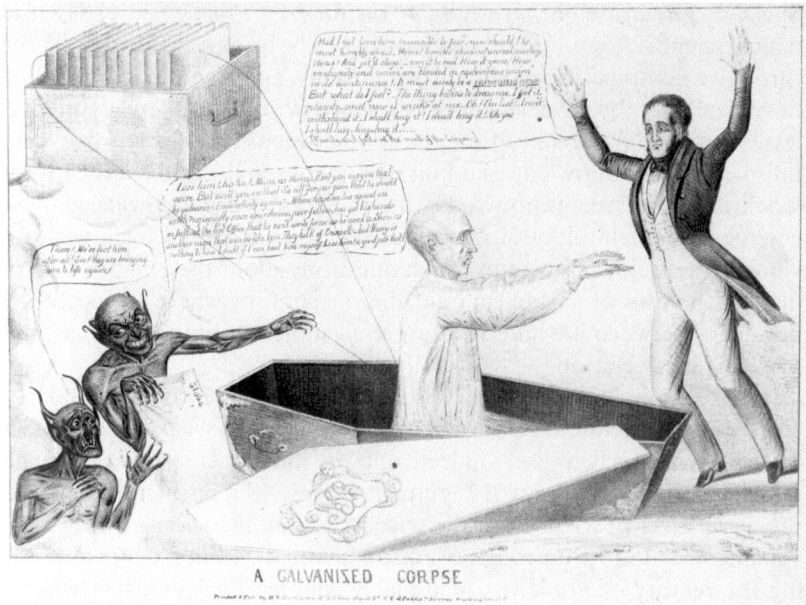

Image 2.1 A galvanized corpse (Printed and published by H.R. Robinson, 1836. http://www.loc.gov/pictures/item/2008661296/. No known restrictions on publication)

His most famous demonstration took place in London, in 1803, when he applied an electrical shock to the cadaver of convict George Foster at the Royal College of Surgeons in London. Professor Nick Groom, of the University of College London, notes that Mary Shelley would have been aware of such experiments, including this "particularly chilling one in London in 1803 when galvanism was used on the body of an executed criminal. The very first thing that happened was that the corpse opened its eyes. A very *Frankenstein* moment."[17]

Popular representations of de-extinction-related research today are replete with just such Frankenstein moments, with references to Dr. Frankenstein's experiment coming in second only to *Jurassic Park* in the mainstream media, as scientists and journalists try to convey both the magnitude of these scientific advances and the limits of their application to popular audiences. Douglas McCauley, an ecologist at the University of

California, Santa Barbara, for example, argues that de-extinction of the woolly mammoth is far enough along the technical trajectory that "it's a bit like finishing the last stiches after creating Frankenstein's Monster—and suddenly we pause for a moment and start to think about whether we should really switch on the power and bring that thing to life."[18] In the article reflecting on the 200th anniversary of Shelley's novel, and its insights for contemporary biological scientists who seek to alter natural processes, C. Nowlin wagers that "how well they will keep Frankenstein's example in mind as they explore unknown genetic frontiers remains to be seen."[19] The persistence of these metaphors and literary tropes serves not only to communicate science to the public but to sublimate our fears and expectations around biotechnology in the twenty-first century. On the one hand, de-extinction research takes place in the most immaculate of scientific spaces. An article on the research of geneticist George Church's team at Harvard University, for example, begins by considering the mammoth cells that sleep "tight in a closet. It's completely dark in there, moist, comfortably warm, with low oxygen content, just like in a uterus. It is frequently monitored; everything in the closet has to be perfect, concerning temperature and air composition."[20] Yet, at the same time, the idea of de-extinction, connecting as it does to cultural fears and aspirations that have persisted through the centuries, tends to break free from these quarantined spaces in a manner that is downright gothic, propelled as it is by the dread of an environmentally bleak future and rife with unanticipated ethical, social, and cultural challenges that complicate the clean official narrative at every turn.

I Just Can't Get That Monster Out of My Mind

Many scientists often express frustration with the *Jurassic Park* metaphor specifically, and with social, artistic, and pop cultural consideration of de-extinction projects generally. In the science journal *Genes*, Ben Novak, a lead researcher on the Passenger Pigeon Revival Project, argues that:

> confusion over de-extinction practice stems largely from the absence of one key voice: de-extinction practitioners. Not one de-extinction paper preceding this manuscript has been authored by a scientist working on the de-extinction breeding programs at the centre of critical attention. The lack of peer-reviewed publications authored by de-extinction program leaders is in part because there are only seven active de-extinction programs globally, most at nascent stages of progress.[21]

While technically correct (though some might quibble about how precisely we can demarcate bona-fide de-extinction practitioners from scientists in other relevant subdisciplines such as molecular paleontology), this observation fascinates on multiple counts. First, even a simple search of the Internet instantly locates thousands of links to mass media articles on de-extinction published since 2013. If Novak is correct, the juxtaposition of the quite preliminary state of the actual science versus the abundance of media commentary on it raises important political and cultural questions about which scientists, fields of inquiry, and institutions have reliable access to the mass media and why. Likewise, Ross McPhee, the curator of the Department of Mammalogy at the American Museum of Natural History, describes de-extinction as "an antiquarian effort, very few people are interested in bringing back a mammoth,"[22] though he is speaking in this context in an article premised on the question of whether or not extinct species should be brought back. Moreover, while the *Genes* article does provide an overview of the state of the art of several in-progress de-extinction projects, it is primarily centered around Novak's proposed clarification of the definition of de-extinction agreed to by the Species Survival Group (IUCN) in 2016 and a defense of the underlying idea. It offers little in the way of demonstrable and verifiable progress toward sustainable de-extinction programs. Finally, one plausible reason for the paucity of peer-reviewed articles about de-extinction could be that much of the de-extinction research announced via press releases and picked up by newspapers around the world references initial promising results that are neither replicated nor confirmed. This also seems indicative of both the type of science that is more likely to be reported and the various professional and material incentives that exist at the science/media nexus to play up that which is provisional and promissory. Many de-extinction advocates use these mass media platforms to put de-extinction in front of the public but then seem to object when non-scientists start to question either the ethics of the proposal or query the actual progress toward promises previously made in the more anticipatory stage of a new endeavor.

Novak's observation is also a textbook example of what sociologist of science Thomas Gieryn refers to as the boundary problem in science and technology studies: "where does science leave off, and society—or technology—begin? Where is the border between science and non-science? Which claims or practices are scientific? Who is a scientist? What *is* science?"[23] Much of the story of de-extinction as real science is inextricably wound up with the deliberate use of the mass media by scientists and entrepreneurs who need to draw attention to their work in a landscape crowded with competing projects, and to the corresponding need of media outlets to guarantee readers and website clicks. Hence, the intrusion of non-scientists, whether the mass public, journalists, or even experts in humanities and social science disciplines who want to engage with philosophical and cultural questions, would seem all but inevitable when proponents of de-extinction launch their work with a highly publicized TEDx conference underwritten by both the National Geographic Society and the Silicon Valley-based non-profit Revive and Restore (affiliated with the Long Now Foundation). Revive and Restore founder Stewart Brand's talk at TEDx in 2013, for example, is billed on the conference website as *The dawn of de-extinction: are you ready*, while Mike Archer, Dean of Science at New South Wales University, gave a related talk entitled *How we'll resurrect the gastric brooding frog, the Tasmanian tiger*. This code switching by de-extinction proponents between expert science, what Evans, Kotchetkova, and Langer refer to as the "constitutive forum," in which only a small number of credentialed experts get to contribute to the debate, and popular science, in which the need to explain bioscience to a wider audience for various reasons means moving into the public arena, ensures slippage between fact and fiction, science and speculation, science and culture. Moreover, since "the contingent forum, which includes everything else that scientists do that is connected with their work, tends to be much more informal, varied and contextual [and] is where most non-scientists will encounter science, the presentation of science within this context is arguably a key influence on perceptions."[24]

The tensions here are embodied in everything from the title of evolutionary molecular biologist Beth Shapiro's well-received book, *How to Clone a Woolly Mammoth*,[25] in which she tells us, essentially, that no one can or ever will be able to (due to the lack of intact cells within even the best preserved mammoth carcasses), to paleontologist John Hawks's excoriation of de-extinction as "fake news." Hawks issued his jeremiad against bad science reporting after stories hit the mass media in February 2017 with headlines such as "Woolly mammoth will be back from extinction in two years, say Harvard scientists"[26] and "Woolly mammoth on the verge of resurrection, scientists reveal."[27] Within days of these initial reports, several media sites had scaled back their claims, noting that the research in question actually involved the use of CRISPR-Cas 9 to insert a few woolly mammoth genes into an Asian elephant genome and that even the production of a hybrid mammophant embryo in the laboratory would still be a long distance from an actual, breathing woolly mammoth in the field. Hawks, for his part, seemed particularly incensed by what he perceived to be casual references to the "artificial womb" in which any mammophant embryo would need to gestate, given the ethical and technical complexities that militate against using an Asian elephant as a surrogate. He argues that artificial wombs are in such a nascent stage of development that the successful use of one would arguably be much bigger news than even a resurrected woolly mammoth. Hawks places the blame on this and other overhyped de-extinction stories squarely on science reporters, arguing that "science clickbait is fake news. Today I count 65 stories, mostly repeating the same stupid mistakes. Scientists need to work together to make journalists step up and do their basic homework on these stories."[28] It is quite probably the case that many of the articles, written on tight deadlines, were either informally crowdsourced or lifted almost verbatim from laboratory and University press releases. That said, Hawks's argument that "what is needed is some basic respect for the facts, and better investigative questions," nevertheless, seems one-sided and again draws a clear bright professional boundary around science, as though these intrusions of the broader culture were somehow unanticipated or unsolicited. As Nik Brown argues, the biotechnology sector "is today synonymous with the language and imagery of futuristic breakthroughs. The whole area is literally spilling over with heated aspirations, promises, expectations, hopes, desires and imaginings."[29] Scientists are often key actors interacting with reporters to promulgate these promises and expectations, to get attention in prestige newspapers, to sell books, underwrite public speaking tours, and to potentially enroll financial backers (Peter Thiel, as one example, donated 100,000 dollars in 2017 to the George Church lab to help fund the woolly mammoth research). De-extinction, however small the actual research community might be, *pace* Novak, commands an extraordinary amount of media and public attention, indeed is the paradigmatic example of promissory twenty-first-century science scaled to the demands of the emergent Anthropocene and to an audience primed to expect ever more spectacular breakthroughs.

Have a Good Anthropocene

The Breakthrough Institute's *Eco-Modernist Manifesto* envisions a "good Anthropocene" in which "humans use their growing social, economic, and technological powers to make life better for people, stabilize the climate, and protect the modern world."[30] Not all eco-modernists are necessarily advocates of de-extinction, but de-extinction as an idea would be impossible to envision in the absence of a deep and abiding faith in the power of science and technology to not only solve the problems of the present and future, but potentially to allow us to rerun the past and nudge it in different directions. Gene therapy, personalized medicine, and de-extinction, while distinct research programs in certain technical ways, converge around a twenty-first-century politics of life in which, according to entrepreneur and scientist J. Craig Venter, "now we can go in the other direction by starting with computerized digital code, designing a new form of life, chemically synthesizing its DNA, and then booting it up to produce the actual organism."[31]

De-extinction is thus a techno-scientific vision fit for the Anthropocene, a term coined by ecologist Eugene Stoermer in 1980 and popularized by Paul Crutzen, a Nobel-winning atmospheric scientist. The basic premise is that "in every respect, the world we inhabit will henceforth be the world we have made."[32] The notion that humankind has become key evolutionary force on this planet can enroll both optimists and pessimists, those who seek to extend our technological prowess and those who would seek, as do deep ecologists, to formulate new ways of relating to the planet that radically de-center the primacy of the human being and deconstruct the edifice of modern technology in order to reduce our environmental footprint. Indeed, as Purdy notes, "talking about 'the Anthropocene' is an attempt to do what the concept of 'the environment' did in the 1960s and early 1970s: join problems as disparate as extinction, sprawl, litter, national-parks policy, and atomic fallout into a single challenge called 'the environmental crisis.'"[33] In this regard, it is a multifaceted concept, albeit one that provides a lens through which to consider seriously whether or not the best solution to the environmental crisis is to ramp up the search for (bio)technological solutions or perhaps to rethink the global economy—and its technological scaffolding—altogether.

Encouraged by several deeply resourced Silicon Valley entrepreneurs and affiliated scientists, de-extinction, in essence, seeks to use advanced biotechnologies such as ancient DNA analysis and reproductive cloning (using closely related surrogate species) to "bring back" certain extinct species, at least in hybrid form, and to retrieve lost genetic information, currently locked up in museum archives, amber, and permafrost, that could possibly revitalize global biodiversity. Though contested within the broad field of conservation science, the underlying idea is no longer quite as fanciful as it may sound and occupies a spot in that liminal space between science fiction and science. In 2003, Italian scientists managed to bring back, in the laboratory, for approximately ten minutes, a cloned bucardo, a notable achievement in that the last surviving member of the subspecies had died in 2000, though fortunately not before skin biopsies had been obtained and frozen. Of course, three-year-old DNA is significantly different in quality from 100-or-1000-year-old DNA, hence the scientific uncertainty that still attends proposals to bring back extinct species such as the Tasmanian tiger or the woolly mammoth. Still, in this experiment, which has yet to be replicated, the extinction barrier was crossed, however briefly.

Indeed, in the realm of modern biotechnology, the seemingly inevitable end of the frontier is always delayed by the discovery of new research avenues, applications, and ideas. Hence, just as the Anthropocene provides one lens for thinking about de-extinction and the relationship between nature and technology, rewilding provides a possible answer to the question of where we would put those revivified woolly mammoths. Rewilding begins from the premise that "Earth is now nowhere pristine, in the sense of being substantially free from human influence,"[34] and then sketches a compelling vision that sets the return of keystone Pleistocene mega-fauna as the temporal benchmark for a world in which vast swathes of habitat are re-connected in North America and evolution is, in essence, reset to travel along its own inscrutable and long-term path. Rewilding does not require de-extinction, as the mega-fauna referred to here are those, such as elephants and cheetahs, that are still extant in pockets of habitat in Africa and parts of Asia, albeit critically endangered. However, in its vision of a restored Pleistocene ecosystem in which large and magnificent animals roam freely across reconstituted forests and plains that stretch beyond the horizon, rewilding taps into both the same human fascination with lost worlds and the promise of eco-redemption that powers much of the de-extinction agenda. Alan Rabinowitz once noted that "the energy in a jungle with big predators is a very, very different energy, and when you truly merge with it and feel it, it's not a dangerous energy. It's not a negative energy—completely the opposite. It's this huge, positive, overwhelming force which humbles you, makes you realize that there are things much greater on Earth than you."[35] It is this urge to reconnect with that primal force that unites, if not the day-to-day science of de-extinction, its underlying rationale and our ongoing fascination with the fictional worlds that writers and filmmakers conjure.

THE ABILITY TO RAISE THE DEAD COULD COME IN HANDY

The desire to harness the power of nature, or even to resuscitate life, threads through modern science and finds expression today in the more ambitious projects of biotechnology companies in Silicon Valley and other global technology centers. Leonid Krasin (1870–1926), a follower of Nikolai Federov (1829–1903) and Russian cosmism (and the man later put in charge of the delicate task of preserving Vladimir Lenin's corpse in a state amenable to potential scientific resurrection), was "certain that the time will come when science will become all-powerful, that it will be able to recreate a deceased organism."[36] In an analysis of the intellectual ructions and even outright despair among many religious believers that Charles Darwin's theories caused in both Britain and Russia, John Gray argues in *The Immortalization Commission* that "during the late nineteenth and early twentieth century science became the vehicle for an assault on death. The power of knowledge was summoned to free humans of their mortality. Science continues to be a channel for magic, the belief that for the human will, empowered by knowledge, nothing is impossible."[37] In that sense, with due respect to the scientists involved, de-extinction at its outer edges enters the sphere of magic: there is something tantalizing about it, but something unsettling as well, which is why it continues to exert a strange fascination despite concerted efforts by some proponents to pull it back from the brink of more spectacular claims and to focus attention on the incremental conservation insights and tools enabled by the approach. De-extinction also provides an opening for public consideration of life-on-demand, that is, the coming conversion of biotechnological research into applied, problem-focused applications suitable to real-world human health and environmental problems. As Narasimhan argues, "the golden era of molecular biology has paved the way for us now to tweak, fix, engineer, or even entirely synthesize genomes at levels and scales never before seen … humans have the power like never before to create, change, or destroy life forms beyond their own kind,"[38] perhaps even the power to bring back some of those Pleistocene mega-fauna (or remarkable genetic proxies thereof).

This brings us back to *Jurassic Park* and that T-Rex. McGregor (2004) argues that "environmental imaginaries are highly contested and can be thought of as the ways in which a society collectively constructs, interprets and communicates nature."[39] Today, the environmental imaginary is ineluctably bio-digital, inseparable from what W. J. T. Mitchells refers to as biocybernetic reproduction in which "the convergence of genetic and computational technologies with new forms of speculative capital has turned cyberspace and biospace (the inner structure of organisms) into frontiers for technical innovation, appropriation, and exploitation—new forms of objecthood and territoriality for a new form of empire."[40] The idea of de-extinction is intrinsically biocybernetic; indeed, while dinosaur DNA and amber got most of the attention in *Jurassic Park*, it is important to recall that it is the failure of the computer programs (and the malfeasance of the computer hacker) that sets all of the chaos in motion. On technical grounds, dinosaur DNA is so old and thus degraded—and the species' natural environment is so out of sync with ours—that the prospect of bringing one back via gene editing or cloning is essentially nil. Yet scientific frustration with social and popular representations of de-extinction misses a larger point about the relationship between science and society in an era saturated with both social media and a fascination with the biotechnological imaginary. Dinosaurs make for particularly potent movie monsters, it is true, but the prospect of being able to generate life on demand did not spring into being with the recombinant DNA research of the 1970s nor with de-extinction research today. This desire tracks back to the very beginning of the modern biological enterprise, as does the science fiction that seeks to tame it, indeed it tracks back in its essential form to our most ancient and sacred myths. To object to *Jurassic Park* because it is not "real," in the sense of being scientifically accurate, is to miss the larger cultural point. These novels and movies channel the anxieties produced by technological advances. They do not need to be technically accurate to highlight very real social concerns attendant upon the unlocking of nature's secrets and the pursuit of technological trajectories that can rapidly develop unanticipated pathways of their own. Scientists cannot conjure with the promise of resurrected life and assiduously promote the idea in the digital media-sphere and then be surprised that the social response does not always respect the boundaries of expert and subdisciplined science. At this juncture, when so much investment capital and technological promise is wrapped up in biotechnology and its ongoing advances, the notion that science can jump readily back and forth between the contingent and the constitutive, but that every other stakeholder in this game must remain corralled on the cultural side of the divide, is as unrealistic in its way as assuming that a dinosaur will stay on its designated side of the fence.

NOTES

1. Moodley, Russo, Robovsky, Dalton, Kotzé, Smith, et al., "Contrasting Evolutionary History, Anthropogenic Declines and Genetic Contact in the Northern and Southern White Rhinoceros (*Ceratotherium simum*)."

2. Bearak, "Sudan, the World's Last Male Northern White Rhino, Has Died, Putting His Species on the Brink of Extinction."

3. Baron, "Inside the Frozen Zoo that Could Bring Extinct Animals Back to Life."

4. Hall, "Resurrecting the Northern White Rhino and Other Species. But at What Cost."

5. Thomson, "Hybrid White-Rhino Embryos Created in Last-Ditch Effort to Stop Extinction."

6. Save the Rhino, "Can We Save the Northern White Rhino."

7. Thomson, "Hybrid White-Rhino Embryos Created in Last-Ditch Effort to Stop Extinction."

8. Sánchez-Bayo and Wyckhuys, "Worldwide Decline of the Entomofauna: A Review of Its Drivers."

9. Hood, "Many Sharks Closer to Extinction Than Feared: Red-List."

10. Scheele, Pasmans, Skerratt, Berger, Martel, Beukema, et al., "Amphibian Fungal Panzootic Causes Satastrophic and Ongoing Loss of Biodiversity."

11. Seddon, "The Ecology of De-extinction," p. 992.

12. Boissoneault, "Jurassic Park's Unlikely Symbiosis with Real-World Science."

13. Cano, Poinar, Pieniazek, Acra, Poinar, "Amplification and Sequencing of DNA from a 120-135-Million-Year-old Weevil."

14. Gutiérrez and Marín, "The Most Ancient DNA Recovered from an Amber Preserved Specimen May Not Be as Ancient as It Seems."

15. Lightman, *Victorian Popularizers of Science: Designing Nature for New Audiences*, 13.

16. Ruston, "The Science of Life and Death in Mary Shelley's Frankenstein."

17. Doward, "Who Put the Spark in Frankenstein's Monster."

18. Heinemann, "Extinction Was Yesterday."

19. Nowlin, "200 Years after Frankenstein," 446.

20. Heinemann, "Extinction Was Yesterday."

21. Novak, "De-Extinction."

22. Frederick, "Should Scientists Bring Back the Woolly Mammoth."

23. Gieryn, "Boundaries of Science," 392.

24. Evans, Kotchetkova, Langer, "Just Around the Corner: Rhetorics of Progress and Promise in Genetic Research," 47.

25. Shapiro, *How to Clone a Woolly Mammoth*.

26. Knapton, "Woolly Mammoths Will Be Back in Two Years, Say Harvard Scientists."
27. Devlin, "Woolly Mammoths on Verge of Resurrection, Scientists Reveal."
28. Hawks, "How Mammoth Cloning Became Fake News."
29. Brown, "Hope Against Hype: Accountability in Biopasts, Presents and Futures."
30. Asafu-Adjaye, Blomqvist, Brand, Brook, Defries, Ellis, et al., *An Ecomodernist Manifesto*, 7.
31. Venter, *Life at the Speed of Light: From the Double Helix to the Dawn of Digital Life*, 6.
32. Purdy, *After Nature: A Politics for the Anthropocene*.
33. Ibid.
34. Donlan, Berger, Bock, Bock, Burney, Estes, et al., "Pleistocene Rewilding: An Optimistic Agenda for Twenty-First Century Conservation."
35. Rabinowitz, "A New Strategy for Saving the World's Wild Big Cats."
36. Tumarkin, "Religion, Bolshevism, and the Origins of the Lenin Cult."
37. Gray, *The Immortalization Commission: Science and the Strange Quest to Cheat Death*.
38. Narasimhan, "Resurrection," 229.
39. McGregor, "Sustainable Development and 'Warm, Fuzzy Feelings': Discourse and Nature within Australian Environmental Imaginaries."
40. Mitchell, *What Do Pictures Want: The Life and Loves of Images*, 309.

CHAPTER 3

The Secrets of All Inheritance: A Cultural History of DNA

Abstract This chapter begins with the award of the 1962 Nobel Prize to Francis Crick, James Watson, and Maurice Wilkins for their work on the structure and function of DNA. It then discusses the emergence of ancient DNA analysis in the mid-1980s, a research trajectory that extended science's reach into the deep past and opened up new ways to leverage the latent power of genetic information. The third section evaluates the legacy of the Thylacine Cloning Project (1999–2005), an initiative of the Australian Museum of Science that introduced the idea of bringing back an extinct species to a broad international audience. The chapter concludes with the question of whether de-extinction falls properly into the domain of science or spectacle.

Keywords Ancient DNA • De-extinction • Tasmanian tiger • Thylacine • Science and spectacle

In 1963, the international edition of *Life* magazine published an article detailing the dismal plight of several mammalian species on the brink of extinction entitled *Africa's animals in peril*. Powerful and disturbing photographs of tiger heads piled up for sale in a shop window in Kenya and a pile of elephant and zebra carcasses carelessly discarded by poachers accompanied the text. The publication of this article dovetailed in time with increasing public awareness of environmental crises and wildlife extinctions

© The Author(s) 2020
A. L. Fletcher, *De-Extinction and the Genomics Revolution*,
https://doi.org/10.1007/978-3-030-25789-7_3

in both the United States and internationally. Rachel Carson's best-seller *Silent Spring* (1962) had recently ignited the mid-century American environmental movement, both because of its compelling presentation of the ravages of extensive use of the synthetic pesticide dichloro-diphenyl-trichloroethane (DDT) in agriculture and the attention shone upon the book when President John Kennedy convened a special panel of the President's Science Advisory Committee to investigate the links between pesticide use, human health, and the environment. Stewart Lee Udall, United States Secretary of the Interior from 1961 to 1969, released *The Quiet Crisis* in 1963, another urgent call to arms for activist environmental preservation that indicted an America "of vanishing beauty, increasing ugliness, of shrinking open space, and of an overall environment that is diminished daily by pollution and noise and blight."[1] Over the next decade, these exposes, in combination with concentrated bipartisan executive and legislative action, produced the *1964 Wilderness Act*, the *1966 Endangered Species Preservation Act*, the *1972 Marine Mammal Protection Act*, and the *1973 Endangered Species Act*. Ten years of negotiation also saw the United States becoming one of the 80 original signatories to the Convention on International Trade in Endangered Species of Wild Fauna and Flora (CITES).

This burgeoning public and political interest in the environmental consequences of industrialization and the use of modern agricultural chemicals occurred alongside significant advances in the field of molecular biology. Indeed, the cover story for the same issue of *Life* magazine that reported on the emerging extinction crisis showcased the 1962 Nobel Prize won jointly by Francis Crick, James Watson, and Maurice Wilkins for "discoveries concerning the molecular structure of nucleic acids and its significance for information transfer in living material."[2] Watson and Crick published two consequential papers in 1953. The first of these described the double-helical structure of DNA and the second revolutionized public understanding of life by positing that "it therefore seems likely that the precise sequence of the bases is the code which carries the genetical information."[3] In its editorial on the implications of Watson and Crick's work, which was intended for a broad, literate, but non-specialist audience, *Life* magazine referred to molecular biology as "the most exciting frontier in all of science today."[4] Given that President Kennedy had only recently launched the Moon Project, this was high praise indeed.

It had taken approximately 70 years to reach this point, dating from Jacques Loeb's instrumental work on embryology and artificial parthenogenesis in sea urchins in the late nineteenth century. Loeb had been among the earliest proponents of the idea that biology could be best thought of as an engineering discipline in which natural organisms were the raw material from which biological tools could be built. In his time, *The New York Times* published accounts of his research under provocative headlines such as "Chemical creation of life" (March 1, 1905) and "Prof. Loeb researches into the true nature of death" (January 1, 1902). Hence, though relatively new to a popular audience, by the time *Life* magazine published its article in 1963, both the "life as code" metaphor and the goal of imposing an engineering framework on biology had already exerted a strong influence on the development of twentieth century science and would prove crucial to the mindset that can conceive of potentially bringing back extinct species through genome sequencing, gene editing, and surrogacy. *Life*'s excited assertion that the DNA is the "amazing chemical control system which governs heredity, and hence all of life on earth"[5] finds a contemporary corollary in biotechnologist J. Craig Venter's declaration in 2013 that "life ultimately consists of DNA-driven biological machines. All living cells run on DNA software, which directs hundreds of thousands of protein robots. We have been digitizing life for decades, since we first figured out how to read the software of life by sequencing DNA."[6]

The next section of this chapter details both how and why the code-script metaphor became so consequential to twentieth century biological research. Putting de-extinction debates within this rich cultural and historical context allows us to trace the emergence of life-on-demand, to comprehend the pathway by which the elegant metaphor of life-as-code provided a framework within which distinct research trajectories in population genetics, embryonic and pluripotent stem cells, genome sequencing, ancient DNA analysis, and gene editing could converge into the disruptive idea of bringing back an extinct species. This next section begins in the early twentieth century, when three scholars independently discovered the work of Gregor Mendel. It proceeds to a discussion of ancient DNA analysis which, beginning in the mid-1980s, conceptually extended the idea of retrieving at least fragments of the code of life well into the deep past, thereby opening up, as scientific knowledge accumulated, new possibilities for manipulating life processes and altering the environmental future. The chapter then considers the Thylacine Cloning Project (1999–2005), an initiative of the Australian Museum of Science in Sydney that introduced the idea of bringing back an extinct species to a broad international audience well before the TEDx De-Extinction event of 2013 and quickly became mired in controversies surrounding both the rationale for and the conservation implications of mucking about with genetic fragments of a species that had already departed the Earth. The conclusion of this chapter considers the bedeviling question of whether seeking to bring back an extinct species falls properly into the domain of science or spectacle.[7]

THE SECRETS OF ALL INHERITANCE

At the dawn of the twentieth century, three researchers, Hugo Devries, Carl Correns, and Erich von Tschermak, independently rediscovered Gregor Mendel's nineteenth century research on hereditary patterns in the transmission of traits in hybrid pea plants. All three botanists had been working on the question of inheritance in plants and found that their work confirmed Mendel's essential findings, which had been detailed in the meticulous scientific notebooks published in the 1860s. In particular, Mendel elucidated the law of segregation, which holds that parents pass both dominant and recessive traits randomly to their offspring, and the law of independent assortment, which holds that traits are passed on independently of other traits from one generation to the next via sexual reproduction. Mendel's work, important on its own merits, also convincingly posited a process through which the advantageous natural adaptations necessary to evolution, as theorized by Charles Darwin in *On the Origin of Species* (1859), actually transmitted from one generation to the next. Though "it took almost three decades before the reconciliation between Mendelian genetics and natural selection was brought about by bio-mathematicians, in particular Ronald A. Fisher, Sewell Wright, and J.B.S. Haldane,"[8] the eventual synthesis of the work of Mendel and Darwin provided the basic foundation for the development of the disciplines of population genetics and molecular biology. Indeed, by 1906, William Bateson had introduced the term genetics to the International Congress of Botany, while Wilhelm Johannsen is credited with the first use of the word gene in 1909. Most crucially, as Evelyn Fox Keller notes, "by mid-century, all remaining doubts about the material reality of the gene were dispelled and the way was cleared for the gene to become the foundational concept capable of unifying all of biology."[9]

Scientific interest in molecular biology accelerated at mid-century, particularly in the United States and Britain, and merged with the complementary post War research emphasis on cybernetics and computing. In 1944, scientists at the Rockefeller Institute confirmed that genes were made up of deoxyribonucleic acid, while in that same year, Oswald Avery (1877–1955) discovered that DNA chemically transmits inheritable traits. Perhaps the most significant event in this timeframe was the publication of physicist Erwin Schrödinger's (1887–1961) book *What Is Life: The Physical Aspect of a Living Cell*, compiled for the lay reader from the transcripts of his 1943 lectures to the Dublin Institute for Advanced Studies. With a startling conceptual leap, Schrödinger proposed that genes either were or contained a code script for life, meaning that it should be possible, ergo, to read the code script of an egg and know "whether the egg would develop, under suitable conditions, into a black cock or into a speckled hen, into a fly or a maize plant, a rhododendron, a beetle, a mouse or a woman."[10] J. Craig Venter acknowledges this as the "first mention of the fact that the genetic code could be as simple as a binary code,"[11] a statement that arguably understates how profoundly the code-script idea would influence the further research trajectory of the biological sciences in the second half of the twentieth century. In confronting "the central problems of biology—heredity and how organisms harness energy to maintain order—from a bold new perspective,"[12] Schrödinger unwittingly anticipates Watson and Crick's eventual discovery of the DNA double-helix, "within whose turns lay the secrets of all inheritance."[13]

However, though the life-as-code metaphor dominated molecular biology for several decades, Schrödinger had actually considered alternatives, noting that "the term code-script is, of course, too narrow. The chromosome structures are at the same time instrumental in bringing about the development they foreshadow. They are law—code and executive power—or, to use another simile, they are architect's plan and builder's craft—in one."[14] The conceptual bridge that connects Schrödinger's thinking in the 1940s and Watson and Crick's eventual use of the code metaphor in the 1950s was the parallel development in that decade of cybernetics and information science.[15] Information transmission would become a key organizing concept in molecular biology largely via the ongoing work of the United States Office of Scientific Research and Development, which had been established by President Franklin D. Roosevelt in 1941 to explore how various sciences could inform weapons development. Within this large bureaucratic structure, Section D-2 of the affiliated National Defense Research Committee, which studied artillery fire and accuracy, was directed by Warren Weaver, a mathematical physicist who coined the term "molecular biology" in 1938. Weaver worked alongside Claude Shannon and Norbert Weiner, who in turn collaborated with John von Neumann, all three of them colossal figures in, respectively, the advancement of information theory, cybernetics, and game theory. Funding from the Rockefeller Institute and the Macy Foundation underwrote significant research progress in information control and feedback, with all of this work culminating in the 1948 Hixon Symposium (published in 1951) wherein von Neumann envisioned the gene as an "'information tape' that could program the organism—like the 'universal Turing machine' described in 1936 by Alan Turing."[16] In his autobiography, molecular biologist Sydney Brenner offers an alternative viewpoint that "the cultural differences between most biologists on the one hand, and physicists and mathematicians on the other" meant that von Neumann's ideas as expressed at the Hixon Symposium did not have an instantaneous impact on the broad field of biology.[17] Nevertheless, von Neumann's comparison of gene function to self-reproducing automata and Watson and Crick's insight that genetic information could be conceptualized as transmitted through the DNA code laid two research tracks that would eventually converge into a broad engineering and information oriented approach to molecular biology. By the 1960s, as American faith in technological progress flourished in the context of both Cold War space exploration and President Kennedy's New Frontier optimism—and advances in computing technology often paced those in biology and exerted a similar strong influence on the popular culture—the code-script metaphor provided a way for mainstream sources such as *Life* magazine to draw analogies that made abstract biological research and concepts somewhat more accessible to an educated lay audience. The code-script made metaphor also presages the eventual emergence of synthetic biology, in which engineers largely invaded the study of genetics, bringing the language of circuit design and software development squarely into the field of biology.

The period immediately following the award of the Nobel Prize to Watson, Crick, and Wilkins saw extraordinary progress in molecular biology and in the ability of scientists to intervene at the molecular level to nudge life processes in specified directions. In 1980, Paul Berg won the Nobel Prize for Chemistry for his research into the biochemistry of nucleic acids and recombinant DNA, sharing the prize with Walter Gilbert and Frederick Sanger for their work in determining the base sequences of nucleic acids. That same year, the United States Supreme Court held, in *Diamond v. Chakrabarty* (447 U.S. 303), that a living microorganism that had been genetically altered in the lab could be patented. The case pertained specifically to a bacterium that had been modified to break down crude oil as a possible sustainable clean-up solution for oil spills. However, the significance of the decision went far beyond this microorganism, extending not only to the field of recombinant DNA, in general, but to the growing financial and business interest in commodifying new biotechnologies and bringing them to market in the energy, health care, and environmental sectors. Patent law soon encompassed not only engineered bacterium but also complex, multicellular organisms. Rudolf Jaenisch and Beatrice Mintz created the first transgenic mouse by injecting foreign DNA into a mouse embryo in the early 1970s. By 1984, a research collaboration between Harvard University and Dupont produced the Oncomouse (also known as the Harvard Mouse), which had been genetically engineered to have an active cancer gene that made it more useful in cancer treatment research. On April 12, 1988, Oncomouse received U.S. Patent 4,736,866, becoming the first animal to be patented in the United States.

Significant research in genome sequencing also took place alongside these advances in patent law and in the production of transgenic organisms. In the 1980s, John Sulston, founding Director of the Sanger Center, who eventually shared the 2002 Nobel Prize for Physiology or Medicine with Sydney Brenner and Robert Horvitz, began work on the nematode worm (*C. elegans*), publishing this first genome sequence of a multicellular organism in 1998 (bacteria, viruses, and yeast had been sequenced earlier). In a discussion of the accomplishment in *Science*, the *C. elegans* Sequencing Consortium noted that "the worm project has contributed to technology and software development; it is not a unique test-bed, but along with the other genome projects, it has explored ways of increasing scale and efficiency" and that work had further direct bearing on the soon-to-be-launched Human Genome Project.[18] Moreover, Sulston's work, as a precursor to the coming era genome sequencing of different species across the globe, was another key moment in which biotechnology continued to entwine with an emergent information society in which the creation, modification, and commodification of information were increasingly the source of wealth.[19]

Finally, in July 1996, scientists at the Roslin Institute took perhaps the most highly publicized step forward, indeed one that "scientific orthodoxy declared...was impossible,"[20] with the birth of Dolly the Sheep, the first mammal to be successfully cloned from an adult cell rather than an embryo. Newspapers around the world carried the announcement of Dolly's birth on February 23, 1997, with the *New York Times* referring to it as "a feat that may be the one bit of genetic engineering that has been anticipated and dreaded more than any other."[21] In the twenty-first century, when biotechnological breakthroughs seem more routine and must be packaged in ever more stupendous stories and glossy images to garner sustained public attention across multiple social media channels, it is important to recall how consequential the Dolly the Sheep announcement was at the time. Lee Silver, a biology professor at Princeton University, for example, argued that "it's unbelievable. It basically means that there are no limits. It means all of science fiction is true. They said it could never be done and now here it is, done before the year 2000."[22] Many newspaper reports focused not on Dolly and her immediate implications for animal breeding, but instead drew links to the controversial and provisional issue of potential human cloning and even the possibility of cryonic resuscitation, via headlines such as "Dolly opens door for life after death" (*The Scotsman*, February 27, 1997), "Hello Dolly, or perhaps Frankenstein" (*The Moscow Times*, February 26, 1997), and "Scientific leap or a monster" (*Canberra Times*, March 1, 1997). The Vatican immediately called for a global ban on human cloning in light of the announcement, while United States President Bill Clinton issued an executive order to ban the use of federal funds for research into cloning human beings and the United Nations began a deliberation process that finally resulted in a 2005 General Assembly declaration (GA10333) requesting that Member States take measures to prohibit all forms of human cloning. Professor Ian Wilmut, the scientist who led the Dolly the Sheep project, took pains in many of his interviews to emphasize that the research primarily concerned efforts to improve livestock productivity and, potentially, human health, and did not seek or intend to conjure with the vexed issue of human cloning. Nevertheless, early critics, such as Dr. Ronald Munson, a medical ethicist at the University of Missouri-St. Louis, anticipated a concern that would later beset proposed de-extinction projects, namely the perception of both scientific hubris and capriciousness: "here we have this incredible technical accomplishment, and what motivated it? The desire for more sheep milk of a certain type." It is, he said, "the theater of the absurd acted out by scientists."[23]

Dolly the Sheep, though a genuinely significant scientific breakthrough that surprised many and captured the attention of the world, did not necessarily imply that the extinction threshold could be crossed. For most mainstream scientists, even in the late 1990s, that prospect remained far outside the boundaries of proper and professional scientific discourse. Yet as genome sequencing, stem cell research, and reproductive cloning continued to advance in the 1980s and 1990s, a parallel research program into the retrieval and analysis of ancient DNA samples also emerged and expanded the scope of how genetic information could be used to inform legitimate scientific questions regarding climate change, human migration, and evolution. The juxtaposition of mammalian cloning and early ancient DNA research reports in both scientific and popular media inevitably brought the question of potentially conquering extinction into the limelight. The next section evaluates the emergence of ancient DNA analysis as a respected approach within the broad fields of molecular biology and population genomics before the chapter moves to the more spectacular dimensions of this young science.

Tunneling Through Time with Ancient DNA

In April 1984, *Technology Review* published an April Fool's Day prank that solemnly announced the birth of two woolly mammoth (*Mammuthus primigenius*) calves in the Soviet Union. The report claimed, tongue-firmly-in-cheek, that an American-Russian scientific collaboration had spectacularly reversed extinction by isolating viable DNA and mammoth ova from a frozen mammoth specimen and then combining it with elephant sperm and an elephant surrogate to produce the first *Elephas pseudotherias*. The journal did eventually publish a retraction, though it is not altogether surprising, in this era of recombinant DNA and gene-manipulated mice, that the hoax managed to snare at least a few readers. More importantly, that same year, a research team led by Russell Higuchi successfully extracted a small amount of DNA from the preserved hide of a quagga (*Equus quagga*), a subspecies of South African zebra that went extinct in the late nineteenth century primarily because of the colonial settlers' desire to remove the animals from potentially cultivatable farmland. The team also sequenced 229 base pairs of quagga mitochondrial DNA. *Chemical Engineering News*, among others, recognized the scientific significance of the work, which showed "for the first time that DNA can be obtained and cloned from the tissue of animals that have been dead for at least 100 years and possibly much longer."[24]

One year later, Svante Pääbo amplified DNA sequences from an Egyptian mummy that was an estimated 2300 years old.[25] The systematic study of ancient DNA, as these fragments came to be called, began to flourish as a research program within the broader rubric of molecular paleontology or paleogenomics. Ancient DNA seemed to provide a powerful new tool for tunneling into the deep past. The realization that ancient DNA could provide new or confirming evidence about migration patterns, climate events, and long-term evolution, and could furthermore be at least partially sequenced, expanded the pool of biological material from which scientific and financial value could be extracted. Because proper ancient DNA analysis seems capable of slowly unraveling mysteries buried in deep time, it also exerts a pull on both the scientific and popular imagination. Krause and Pääbo, for example, both expert scientists in this domain who work well within the boundaries of credentialed science, argue that genetics is a form of historical analysis in which the ability to analyze ancient DNA allows us to overcome the "time trap" and to, metaphorically, see the past in ways that were not available to us in earlier eras.[26] The ability to tunnel ever deeper into the molecular past depends upon steady decreases in the cost of genome sequencing and complementary improvements in sequencing technology. Svante Pääbo, for example, notes that "ancient DNA research, defined broadly as the retrieval of DNA sequences from museum specimens, archaeological finds, fossil remains and other unusual sources of DNA, only really became feasible with the advent of techniques for the enzymatic amplification of specific DNA."[27] The Sanger sequencing method, the only large-scale method available for approximately 30 years, could not provide useful results on fragmented and contaminated ancient DNA samples, effectively ruling out comprehensive genomic analysis of extinct species. The debut of the 454 method in 2005, however, increased sequencing capacity by two to three orders of magnitude, with a throughput of 20 million base pairs per run. Subsequent improvements to the 454 method, as well as parallel developments such as Solexa, SOLiD, and Heliscope, could efficiently generate ten billion base pairs of sequence data per run and produce relatively short sequence reads that were particularly suited to analysis of ancient DNA samples of less than 100 base pairs.

Nevertheless, scientists who pioneered the use of ancient DNA as an analytical tool often struggled to keep the public and mass media focused on the published and incremental findings of the work as opposed to the murmurings from the *Jurassic Park* corner and endeavored to produce universal criteria for ancient DNA analysis that included, centrally, the posing of a legitimate research question derived from the existing literature and an acute sensitivity to the contamination and replication challenges inherent in the discipline. Yet by December 1997, only a few months after Dolly made her world debut, a newspaper article entitled "Reviving the ice age in an era of global warming" not only anticipated one of the key justifications that would be made in support of woolly mammoth resurrection proposals 20 years later (the need to bring back the woolly mammoth, even in hybrid form, so that the renewed species could play again a keystone role in maintaining the permafrost and thus stabilizing climate emissions), but also reported on the work of Japanese geneticists Kazufumi Goto and Shoji Okutsu, who expressed hope that they could eventually extract preserved DNA in good condition from a perma-frosted mammoth and then combine mammoth sperm with an elephant egg to produce a half-mammoth/half-elephant hybrid. In October 1999, a French-led research team, partially funded by the Discovery Channel, extracted the world's first reported intact woolly mammoth carcass (though subsequent corrections noted that the mammoth's head had to be extracted separately), estimated to be approximately 20,000 years old, from the Siberian permafrost. The 23-ton Jarkov Mammoth ignited media speculation about the possibility of recovering sperm or DNA from the specimen that could then be used to clone a mammoth. Though paleontologist Dick Mol, one of the researchers, emphasized how ancient DNA analysis could help to determine the exact reason the species died out, even the elite *Washington Post* could not resist headlining its article "In the tundra, a giant genetic find; fully preserved woolly mammoth may be reproduced" (October 21, 1999). It is against this backdrop of media interest, the worldwide success of the Jurassic Park franchise, and accelerating progress in biotechnological research that the Australian Museum in Sydney launched what would become a cautionary tale for anyone who sought to engage in debate about the possible synergy between reproductive biotechnologies, ancient DNA, and the problem of extinction.

SEE IT NOW, WHILE IT'S STILL EXTINCT

On September 7, 1936, the last known extant Tasmanian tiger died in captivity at Beaumaris Zoo in Hobart, Tasmania, due to human neglect that resulted in its exposure to unseasonably cold weather. The last thylacine was so unremarked upon that her keeper forgot that afternoon to herd the animal inside its enclosure to protect it from a cold snap. Indeed, there was so little regard for the species at this juncture, whose numbers had been thinned precipitously in the colonial period due to a deliberate campaign to exterminate it, that this particular specimen was finally tossed away in the garbage. At the turn of the twentieth century, the species had a potent (and probably inaccurate) reputation as a rapacious devourer of sheep, which ultimately pitted an already depleted thylacine population against the political and technological might of the Van Diemen's Land Company, farmers, and the Government.

As sad as this story is, the Tasmanian tiger also bridges a key historical juncture between the slow end of the colonial era and the nascent emergence of modern Australia. Though formal recognition of the thylacine's critically endangered status occurred much too late to make a material difference to its survival, early conservationists had in fact added it to a list of protected species drawn up and approved by the Government in late July 1936. The fact that the thylacine slipped so unremarked through this two-month gap between the original settlers' negative attitudes toward indigenous Australian fauna and an emerging preservationist ethos ensures its iconic status, while its taxonomic distinctiveness ensures its relevance to the de-extinction issue today. Perhaps fittingly, the last days of the thylacine also ran roughly parallel to the development of more portable and user-friendly hand-held movie cameras. Ergo, in addition to the photographs stored in official archives, a short reel of black and white footage of one thylacine pacing its cage and resting in the sun is readily available today on the Internet.[28] There is something simultaneously moving and eerie about the flickering image of this animal, suspended in celluloid purgatory, one of the very last survivors of a species that made its initial appearance on Earth some four million years ago (Image 3.1).

In September 1999, the Australian Museum announced the launch of a research project that would attempt to clone the Tasmanian tiger, to be conducted under the direction of then Director Mike Archer (later a Professor of Paleontology at the School of Biological, Earth, and Environmental Sciences, University of New South Wales). Over its lifespan (1999–2005), the project raised an estimated 300,000 Australian dollars from private donors. Upon shutting down the project in February 2005, the Museum, under a new director, released a press statement noting that the available thylacine DNA was too degraded to construct a DNA library, a necessary step before cell cultures could even be considered. During the course of the project, a substantial amount of negative press coverage emanated from both expert and non-expert observers. Janette Norman, the senior curator of molecular biology at Museum Victoria, for example, castigated the project as

Image 3.1   Photograph taken in 1936 of a Tasmanian tiger at Beaumaris Zoo, by Ben Sheppard (Permission of the State Library and Archive Service, Libraries Tasmania, Image NS1298-1-1880_03)

"outrageous," arguing that "you're looking at five Nobel prizes to overcome what we don't know about."[29]

Moreover, the popular perception that biotechnological breakthroughs now provided a potential means to conquer extinction injected a tricky ethical dimension into the broader debate about the project. Even if the project succeeded, a remote possibility that would have taken us to the outer boundaries of science at that time, the ancient DNA samples would have been drawn from archived specimens that dated to approximately 1866 and which were thus relatively well preserved. These samples ultimately proved to be degraded beyond repair and even in the best of circumstances would not have revealed much about the possibility of bringing back extinct species, such as the woolly mammoth, which had died out considerably earlier, nor about how sufficient genetic diversity to enable a sustainable and self-reproducing thylacine population could be generated. Finally, even had the project reached fruition, the seemingly intractable problems of habitat depletion and illegal trade in endangered species would remain.

In its defense, the thylacine cloning project introduced a large and international audience to the new science of ancient DNA and raised the visibility of both the extinction crisis and the state-of-the-art in reproductive biotechnology. However, the mass media's insistence that "scientists are eagerly searching their specimens for material with which to explore the possibilities of bringing some of these creatures back from the dead,"[30] ensured that the thylacine cloning project, whatever the Museum's original intentions, ricocheted between the need to promote the research to encourage both additional funding and audience numbers and the demands of science wherein the breakthroughs announced to the mass media were immediately subject to intense and often harsh professional and ethical scrutiny. Yet the boundary between science and spectacle, in this case, was perhaps not as stable as some stakeholders preferred to believe. While the specific idea of bringing back the thylacine using advanced biotechnology was innovative and guaranteed to provoke a reaction, the lure of conjuring with the vital past can be traced back much farther, making the Tasmanian tiger one key player in a drama that includes earlier stars such as mummies and aurochs.

I Wish to Try to Resuscitate a Mummy

In 1827, Jane Webb-Loudon published *The Mummy! A Tale of the Twenty-Second Century*. The novel borrows rather openly from the basic structure of *Frankenstein* but despite the somewhat overwrought passages scattered throughout, the story speeds along with the narrative verve of a good potboiler. The basic plot concerns Mr. Edric Edmund, a second son who seeks to distinguish himself in science and overcomes his initial squeamishness about touching the dead to travel to the great pyramid of Cheops, there to re-animate the ancient monarch and solve the mystery of life after death. Of course, his wish to try to resuscitate a mummy wreaks all sorts of scary havoc that expands outward to include his family, his romantic interest, and even England. It is also noteworthy that all of this action takes place in the year 2126.

Though more populist than Mary Shelley's iconic work, Webb-Loudon's *The Mummy* likewise reflects that tumultuous era in which science and spectacle began to collide and rigid boundaries between the two spheres had not yet been drawn. Kathleen Sheppard, historian of science and technology, has done significant work on the cultural impact of mummies, noting that public displays of mummified remains date to the sixteenth century, while in the nineteenth and early twentieth centuries mummy unwrappings could readily shift "between public spectacles which displayed and objectified exotic artifacts, and scientific investigations which sought to reveal medical and historical information."[31] Mummies, of course, also bring with them the mystery and allure of ancient Egyptian beliefs regarding the immortal soul, a worldview in which death, in an elegant turn of phrase by the Canadian Museum of History, is but a "temporary interruption."[32] From a practical point of view, the methods of mummification helped to ensure that some mummy specimens would provide valuable sources of ancient DNA whose analysis has, in turn, strengthened the scientific legitimacy of paleogenomics.

Essentially from the start, then, the multiple sciences historically relevant to de-extinction today have often shifted between the public square and the citadel, which matters because it casts some of the current angst about de-extinction research and how it is presented in the mass media in a different light. Jean Baudrillard noted in *Simulations* in the mid-1980s that "science fiction … is no longer anywhere, and it is everywhere, in the circulation of models, here and now, in the very principle of the surrounding simulation."[33] This is arguably a more elegant way of expressing what P. T. Barnum had already figured out in 1866, namely that "there is scientific humbug just as large as any other."[34] The edifice of science depends upon a set of professional norms and verification procedures that require extensive effort to master and which are necessary to ensure reliable results. However, in the case of de-extinction, there is so much cultural baggage attendant upon the idea of reanimating life, which some proponents do draw upon selectively at times to engage with the public, that slamming the door shut on non-expert speculation at this stage would seem ill-advised and also impracticable. Moreover, even within science proper, the thylacine cloning project had an important predecessor in the work of German biologist Heinz Heck, former director of the Berlin Zoo, who began to back-breed the auroch during the 1930s and who resumed the work following the end of World War II. Heck, an impeccable scientist, nevertheless slips into that narrow gap between hubris and hope when he writes, "I myself was curious to know what these animals looked like today. Another reason for doing it lay in the thought that as man cannot be halted in his mad rage for destruction of himself and all these other creatures, it is at least a consolation if some of those kinds of animals he has already exterminated can be brought to life again."[35]

NOTES

1. Udall, *The Quiet Crisis*, 1.
2. https://www.nobelprize.org/prizes/medicine/1962/summary/.
3. Watson, Crick, "Genetical Implications of the Structure of Deoxyribonucleic Acid."
4. Hills, Rosenfeld, "Nearer Now: Control of Aging and Heredity," 45.
5. Life International, "DNA's Code: Key to All Life," 40.
6. Venter, *Life at the Speed of Light: From the Double Helix to the Dawn of Digital Life*, 6.
7. See Fletcher, "Genuine Fakes: Cloning Extinct Species as Science and Spectacle."
8. Diechmann, "Gemmules and Elements: On Darwin's and Mendel's Concepts and Methods in Heredity."
9. Keller, *The Century of the Gene*, 3.
10. Schrödinger, *What Is Life*.
11. Venter, 3.
12. Ibid.
13. Venter, 4.
14. Schrödinger.
15. Cobb, "When Genes Became Information," 504.
16. Ibid.
17. Brenner, *My Life*, 36.
18. C. elegans Sequencing Consortium, "Genome sequencing of the Nematode *C. elegans*: A Platform for Investigating Biology."
19. Garcia-Sancho, "Mapping and Sequencing Information: The Social Context for the Genomics Revolution," 18.
20. Williams, "Death of Dolly Marks Cloning Milestone," PR209.
21. Kolata, "Scientists Report First Cloning Ever of Adult Mammal."
22. Ibid.
23. Ibid.
24. Chemical Engineering News, "Gene Fragments from Extinct Animal Cloned," 8.
25. Pääbo, "Molecular Cloning of Ancient Egyptian Mummy DNA."
26. Krause, Pääbo, "Genetic Time Travel," 9.
27. Pääbo, Poinar, Serre, Janeicke-Despres, Hebler, Rohland, et al., "Genetic Analyses from Ancient DNA," 645.
28. See https://aso.gov.au/titles/historical/tasmanian-tiger-footage/clip1/. Accessed April 23, 2019.
29. Agence-France Presse, "Bid to Clone Extinct 'Tasmanian Tiger' Assailed."
30. The Sydney Morning Herald, "Tassie Tiger Pup that Holds the Hope for Revival."

31. Sheppard, "Between Spectacle and Science: Margaret Murray and the Tomb of the Two Brothers," 525.
32. https://www.historymuseum.ca/cmc/exhibitions/civil/egypt/egcr04e.html.
33. Baudrillard, *Simulations.*
34. Barnum, *The Humbugs of the World.*
35. Heck, "The Breeding-Back of the Aurochs," 122.

CHAPTER 4

Mammoths, Museums, and Molecules: A De-Extinction Icon Emerges

Abstract Woolly mammoths are flexible creatures, able to move even in extinction between science and spectacle, the deep past, present, and future, and between the spheres of the living and the dead. This chapter traces how the woolly mammoth has interacted with science and entertainment throughout the modern period. It first places the mammoth in the cultural context of the nineteenth-century museum and then reflects on Lyuba, a mammoth specimen that holds a special fascination for audiences and illustrates the mammoth's temporal and cultural pliancy. The third section discusses the woolly mammoth's emergence as a de-extinction icon that holds out promissory hope for nature. The chapter concludes with a discussion of life-on-demand and the meaning of a mammoth in a time out of joint.

Keywords Anthropocene • De-extinction • Museums • Natural history • Victorian era • Woolly mammoths

The woolly mammoth (*Mammuthus primigenius*) has been extinct for approximately 10,000 years, though a small, isolated population managed to survive on Wrangel Island, in the far reaches of the Arctic Ocean, finally dying out some 4000 years ago. Our collective memory of the species never entirely vanished, intruding periodically into our awareness like fragments from a dream extending through the centuries. Rouffignac Cave, in

© The Author(s) 2020
A. L. Fletcher, *De-Extinction and the Genomics Revolution*,
https://doi.org/10.1007/978-3-030-25789-7_4

the Dordogne in France, is also known as *les grottes aux cents mammouths* (the Cave of a Hundred Mammoths), as it contains numerous Paleolithic cave drawings of mammoths (alongside bison, the woolly rhinoceros, and ibex) estimated to be 13,000 years old. These images suggest that our ancestors considered the woolly mammoth, valued it, and felt the need to represent it for reasons about which we can only speculate. One elegant scientific theory, for example, postulates that cave paintings, which are often found in difficult to reach areas of the cave that are also acoustic hot spots with strong echoes, were crucial to the human development of symbolic thinking.[1] If this is correct, one person's desire to depict the woolly mammoth to the larger group led to the earliest development of sounds to name and explain it and thus helped our predecessors begin to speak. Moreover, the oldest known artistic representation of a human being, a woman's face meticulously carved into a piece of mammoth ivory, is approximately 26,000 years old,[2] and also connects us through the mammoth to our prehistoric past.

Mammoth fossils, ivory, and permafrosted remains carry deep time into the present. Progress in ancient DNA research extends this historical and scientific gaze to the molecular level and "has provided direct realization of the age-old and immensely seductive idea that nature speaks, and astonishingly enough, it speaks to us in the simple code of nucleotide sequences."[3] The intricacies and riddles of the deep past, fixed and, until now, largely inscrutable, become legible as ancient DNA analysis of fragmented samples allows experts to test theories about human migration, species evolution, and climate change. The promissory aspect of ancient DNA also finds expression in the notion of de-extinction, which renders natural and seemingly finite processes such as extinction more fluid and malleable. DeLord argues that "extinction cannot be reduced to the simple death of individuals; it also has to account for the end of the recognition process and of the transmission of information."[4] The idea of de-extinction, which is reliant upon both reproductive biotechnologies and twentieth-century progress in ancient DNA analysis, is explicitly about the information transmission function of genes, seeking as it does to reintroduce lost DNA into both present and future nature. Even if one takes the more delimited definition of de-extinction as genetic rescue, in which valuable genetic information (though not necessarily full species) will be reintegrated into the global genome, the linear progression from life to death now seems less absolute than it did before and our ability to mix and match genes more powerful.

Since the mid-1990s, the woolly mammoth has occupied center stage in both popular and scientific accounts of de-extinction, which seeks to bring back extinct species and/or to introduce lost genetic sequences into extant genomes (producing, e.g., a "mammophant") via the use of tools such as back-breeding, genome editing, reproductive cloning, and ancient DNA analysis. De-extinction as a media event and a disruptive idea has in the last 10 years moved from the fringes of legitimate science to the mainstream, due to rapid advances in biotechnology, the financial support and coordination of organizations such as the Revive & Restore Project (based in Silicon Valley and affiliated with the Long Now Foundation) and public interest in the possible return of extinct charismatic species. Indeed, the woolly mammoth, that "archetype of everything icy and Paleolithic,"[5] is often enrolled as a particularly promising candidate for eventual resurrection, due to the availability of relatively intact remains preserved in Siberian permafrost and because it has a close extant relative in the Asian elephant, which could potentially serve as a surrogate. Despite the efforts of many scientific experts to emphasize the basic science that can be properly conducted with the tools of ancient DNA analysis, mass media coverage of bio-imaginary resurrected woolly mammoths suggests that many of us cannot help but dream of magnificent herds rumbling across bleak, cold landscapes again. Numerous scientific platforms designed to showcase progress in de-extinction-related research also skillfully deploy the woolly mammoth's legend and image, thus, further smearing the clear bright line that some stakeholders try to draw between mammoth facts and mammoth fantasies.

The idea of the reanimated woolly mammoth provides a means for those outside elite scientific circles to conceptualize and engage with the emerging world of "point-and-click biology" (Incyte Pharmaceuticals). An actual resurrected woolly mammoth would also have a dual status as both animal and machine, as wild creature and (possibly) patentable object, raising complex issues of ownership, commodification, and governance (via the *Endangered Species Act*, any woolly mammoth or mammoth-elephant hybrid released in the United States would immediately be classified as endangered and subject to significant regulatory oversight, while the CITES convention could complicate efforts to move one across national borders). It would thus be a prime example of a new bio-object, a techno-scientifically produced biological entity in which "the boundaries between human and animal, organic and nonorganic, living and the suspension of living (and the meaning of death itself), are often questioned and destabilized, and their identities have to be negotiated and (temporarily) stabilized, and so given an identity."[6] While it may, indeed, be time to put the dead to work,[7] doing so raises intricate scientific, ethical, and cultural questions about the limits, if any, we might put around our technological ambitions and the kind of nature and species we hope to produce.

Finally, the woolly mammoth's image draws the public's attention to books, conferences, magazines, and websites, becoming a promissory object through which multiple anticipatory claims about both our ability to manipulate life through biotechnology and the techno-revitalization of nature translate into media spectacles, investment capital, and hope. The same anticipatory imperative that drives research and investment in the medical sector has now taken hold in the broad field of conservation as some stakeholders mobilize the promise of genetic rescue to garner support and attention. The woolly mammoth thus becomes a key lens through "which the future is being vigorously imagined."[8] The next section of this chapter evaluates the significant role of woolly mammoths in the early history of natural science museums, focusing on the Victorian era and the way in which mammoths and mastodons were used to illustrate new and controversial ideas about deep time and extinction. The chapter then analyzes Lyuba, an extraordinary woolly mammoth specimen excavated in 2007, in the context of both critical concerns about the Anthropocene and the countervailing ecomodernist narrative of de-extinction. As will be argued, Lyuba functions as a crucial boundary object in the science and society nexus, one capable of mediating across different agendas and multiple stakeholder domains and reconciling the environmental past with the promissory future.

THE VICTORIAN GAZE: MAMMOTHS IN THE MUSEUM

Museums, as Steven Lubar argues, "supply both the authentic presence of the artifacts from the past and the possibility of alternative ways to understand that past. Objects don't have fixed meanings. Collected for one purpose, they are used for many purposes. They are open to endless reinterpretation, a resilience invaluable both for individuals and communities."[9] By these lights, famous and highly publicized woolly mammoth specimens such as Jarkov and Lyuba are recent entries in a long line of predecessors that tracks back to the Victorian era and have been subject to interpretation and reinterpretation as science, activism, and entertainment coevolved throughout the modern period. Charles Darwin published *On the Origin of Species by Means of Natural Selection* in November 1859. His work built upon an Enlightenment scientific tradition that included previous gentleman scholars as Jean Baptiste Lamarck (1744–1829) and Alexander von Humboldt (1769–1859). Darwin's particular genius was in synthesizing various strands of the geological and biological science debates of his era into a clear and compelling explanation of how evolutionary change occurs across time. As is well known, the publication of the book galvanized the scientific community and introduced a relatively broad literate public to scientific findings that would eventually undermine a serene belief in a literal interpretation of the Bible and in a stepladder hierarchy of species that invariably put man, by rights, at the top.

It is against this backdrop that the woolly mammoth and the mastodon began to emerge as key artifacts in the translation of scientific ideas into public museum displays. Georges Cuvier (1769–1832), who was instrumental to the early development of the National Museum in Paris and have a significant influence on Darwin, published a scientific paper in 1800 (having previously read it to the scientific community in 1796) that demonstrated how the lower jawbone of the extant Indian elephant was different to that of fossilized mammoths, thus providing support for the then controversial theory that different species had previously existed on Earth and gone extinct.[10] (Cuvier, unlike Darwin, accepted catastrophism, the theory that life forms had been shaped on Earth through sudden and violent natural events that occurred periodically across time.) Just as nucleotide sequences, per Evelyn Fox Keller, seem to speak to us today, fossil analysis opened up new scientific and narrative possibilities for Victorian and Edwardian audiences. As the *Encyclopedia Americana* of 1920 notes:

> animals of long past ages may be preserved unchanged wholly or in part by a fortunate accident. The Siberian mammoths and woolly rhinoceroses frozen into perpetual ice, and the insects protected from decay in the Baltic amber and other ancient resins, are good examples ... The study of fossils (palaeozoology, palaeobotany) has carried our knowledge of the life of the globe back to its beginning, and illustrated by actual examples the steps in its evolution.[11]

The intense scientific debates about evolution and extinction furthered by these fossil remains converged with and facilitated the emergence of the modern natural history museum. For example, the American artist and entrepreneur Charles Wilson Peale, who opened the first public natural history museum in the United States in 1786, was inspired to so do by the discovery of the fossils of a mastodon, an evolutionary cousin of the woolly mammoth that emerged approximately 30 million years ago and also went extinct during the Pleistocene era. The subsequent spectacular displays of models and reconstructions of such ancient beasts brilliantly reconciled the growing need to bring in paying museum audience while also normalizing new ideas about deep time that cast the story of the Earth (and by association, humankind) in a fascinating new light. As Richard Conniff argues "though dinosaurs now come to mind when we think about lost worlds, mammoths and mastodons provided the first persuasive evidence that one of God's creatures could go extinct."[12]

Ten years before Peale opened his museum, inventor James Watt (1736–1819) debuted his significant design improvements to the steam engine, the invention that arguably launches the slow but inexorable shift from the Holocene to the Anthropocene.[13] As early as 1907, the French philosopher Henry Bergson recognized the epochal influence of this device:

> In thousands of years, when, seen from the distance, only the broad lines of the present age will still be visible, our wars and our revolutions will count for little, even supposing they are remembered at all; but the steam engine, and the procession of inventions of every kind that accompanied it, will perhaps be spoken of as we speak of the bronze or of the chipped stone of pre-historic times: it will serve to define an age.[14]

This corresponding fact bears mention in that, while the woolly mammoth embodied and translated deep time in the Victorian era, today specimens like Lyuba are routinely deployed to narrate the Anthropocene, a conceptual era (which proponents consider geologically distinct from the Holocene) that foregrounds the ineluctable and damaging influence on the global ecosystem of humankind and its intensively carbon-based activities. Ironically, the rapid melting of the permafrost, which has recently revealed so many long-buried mammoth specimens, is itself attributable in large measure to anthropocentric global warming. The next section takes up the issue of woolly mammoth as boundary object in the natural history museums of today, where both the reasons for the species' extinction and what it might portend for our own fate resonate throughout the culture.

FROM PLEISTOCENE TO ANTHROPOCENE: LYUBA AND HER KIND ACROSS TIME

In this twenty-first-century biotechnological age, as the pace of progress accelerates (though practical applications often seem to lag), woolly mammoths not only retain their Pleistocene identity but can also be used in contemporary museum exhibits, newspaper articles, books, and documentaries as both sentinels of our present ecological crisis and as promissory figures of a future nature made viable again via biotechnology and human ingenuity. As the Anthropocene bears witness to the destruction of arable land, the degradation and intermittent scarcity of water, rising oceans and temperatures, and an extinction crisis so profound that it may signal an ongoing sixth mass extinction event,[15] there is both knowledge to be gained and hope to be inspired by returning to these mammoth specimens. Lyuba is an exceptionally well-preserved 42,000-year-old woolly mammoth (*Mammuthus primigenius*) calf specimen, discovered in 2007, whose likeness and images travel the world as a key attraction in natural history museum exhibits devoted to the Ice Age.[16] Her arrival in a global city tends to generate enthusiastic reviews, ticket buyers, and media interest. When Lyuba arrived at the Australian Museum in Sydney as part of the *Mammoths of the Age Ice Age* exhibit in 2017, the *Sydney Morning Herald* noted that this was only the fifth time that the Shemanovksy Museum in Russia had released the actual specimen. The *Chicago Tribune* (May 27, 2015) reported that Lyuba's handlers had to schedule a repeat engagement with the Field Museum as the initial exhibit had not been able to satisfy popular demand.

The Lyuba specimen performs important cultural work as a museum boundary object that can time travel between the Pleistocene and the Anthropocene and back again. Boundary objects are defined in science and technology studies as "objects which are both plastic enough to adapt to local needs and the constraints of the several parties employing them, yet robust enough to maintain a common identity across sites."[17] As a boundary object, Lyuba draws together the myriad interests of scientists, philanthropists, entrepreneurs, curators, journalists, natural history museums, the National Geographic Society, and the general public in the co-construction of a compelling long-scale geological narrative that can be extended into the far future. Originally found by a young Nenet reindeer herder in 2007, Lyuba also implicitly connects global audience with the small, indigenous population of the Yamal Peninsula in the Arctic Circle, who believe that this frozen mammoth calf, approximately 1 month old at death, escaped from the world of the dead to haunt the living.[18] Lyuba is ultimately robust, flexible, and interesting enough to bridge the mysteries of the deep past to our present extinction crisis and then to raise the promise of environmental redemption through biotechnological innovation. As Fletcher argues, "from the moment it walked out of the Ice Age and onto the world stage, the mammoth has been pulled incessantly between the demands of science and the lure of spectacle, with spectacle usually winning."[19] Lyuba's identity as a woolly mammoth calf of the Pleistocene era remains fixed, but she—and by relation, her species—can be framed and narrated in myriad ways that have served the often conflictual interests of the natural history museum, science, and publicists for over 150 years.

Lyuba, as both a stand in for her species and as a boundary object, can also illustrate the anthropocentric narrative as effectively as she embodies the enduring mysteries and fascinations of the past. Bernard Buigues, vice president of the Geneva-based International Mammoth Committee, argues that "Lyuba's discovery [in 2007] is an historic event. It could tell us why this species didn't survive … and shed light on the fate of human beings."[20] The three dominant and rival scientific theories of the woolly mammoth's demise are: (1) a severe infectious disease outbreak; (2) over-hunting by early humans; or (3) an extreme climactic event. These theories have an uncanny resemblance to the fears that haunt the contemporary mind such as global zoonotic epidemics, large-scale nuclear war, or catastrophic climate change. Indeed, the same natural history museum exhibits that feature Lyuba often include (as did the San Diego Natural History Museum in 2013) an I-Max 3-D movie feature entitled *Titans of the Ice Age*. This sophisticated digital movie takes the viewer on a virtual trip through a Pleistocene tundra filled with herds of woolly mammoths and ends with a reflection on the need for us to act urgently to forestall further environmental destruction (with a particular emphasis on the mitigation and reduction of climate emissions). The woolly mammoth thus functions as an instantly recognizable media icon through which scientists, activists, the mass media, and the general public can each contextualize the journey from the Pleistocene to the Holocene to the Anthropocene and put the abstract concept of extinction into tangible form. One sample news account of Lyuba's appearance at the Science Museum in Tokyo in 2008 crystallizes the appeal of this specimen to a range of disparate and but interconnected museum stakeholders and participants:

> Meanwhile, at a display in central Tokyo, children peered into a freezer displaying Lyuba's shriveled body. The mammoth is on display until late

February. "It looked amazing, almost like it was alive," said 10-year-old Chikara Shimizu. "Maybe they found Lyuba because the ice in Siberia is melting from global warming," said Chikara's father, Misao Shimizu. "I find that very worrying."[21]

The dystopian Anthropocene vision of a climate out of control operates alongside a promissory counter-narrative wherein our environmental redemption will occur in the intelligent application of new technologies such as geo-engineering, artificial intelligence, and advanced biotechnology. Lyuba, so often deployed as a cautionary figure, is buoyant enough to be, when needed, a symbol of hope for the prospects of de-extinction. Moreover, because the actual Lyuba remains are, to date, the best preserved yet found in the Arctic, she is often used to tease the possibility of eventual mammoth resurrection. In 2009, a *National Geographic* documentary explicitly raises the possibility of using DNA extracted from Lyuba to clone a woolly mammoth, though finally backs away from endorsing the idea, quoting paleontologist Dan Fisher to the effect that, "'cloning an animal as complex as a mammoth is far beyond our current technical capabilities, but there has been remarkable progress on various aspects of the problem. One day perhaps.'"[22] The ability of Lyuba, in particular, and the woolly mammoth, in general, to take on a promissory dimension with respect to future nature helps to keep the audience both intrigued by the ongoing evolutionary saga and attuned to media reports of de-extinction progress that seem to move us, however infinitesimally, closer to the finish line of mammoth resurrected. As Professor George Church speculates "maybe we can no longer delay death, but we can reverse it."[23]

RECIPE FOR A RESURRECTION

As noted in Chap. 2, speculation about cloning a woolly mammoth erupted at the turn of the twenty-first century, contemporaneously with the arrival of Dolly the Sheep, new advances in ancient DNA analysis, and an intense media focus on mammalian cloning. In 1996, Japanese scientists Kazufumi Goto and Shoji Okutsu declared their intent to secure an intact mammoth cell and produce an elephant-mammoth hybrid in the lab over the next several decades. Three years later, the first sequences of Pleistocene-era nuclear DNA, extracted from Alaskan and Siberian mammoth specimens, a 13,000-year-old ground sloth specimen, and a 33,000-year-old cave bear specimen were published.[24] In 1999, that same year, the discovery of a well-preserved mammoth specimen named Jarkov generated global media speculation about eventual cloning and headlines such as "Can cloning restore pitter patter of mammoth reet" (*The Palm Beach Post*), "Goodbye Dolly, hello hairy, prehistoric mammoth clone" (*The Sunday Herald*), "Scientists hope to clone pachyderm" (*The Washington Times*), and "Bringing the mammoth back to life" (*The Times*). By 2000, the woolly mammoth was "one of the best characterized extinct animals at the biomolecular level."[25] The Discovery Channel dedicated two documentaries to the extraction of Jarkov from the ice and the subsequent scientific analysis of the specimen, *Raising the Mammoth* (2000) and *Land of the Mammoth* (2000). By 2006, two major research projects reported results, respectively, of the first complete mitochondrial DNA sequence obtained from an extinct animal (the woolly mammoth)[26] and the sequencing of 13 million base pairs of nuclear DNA taken from a woolly mammoth bone.[27] Finally, in a highly publicized step forward in understanding mammoth evolution and extinction, the Mammoth Genome Project (Pennsylvania State University), under the leadership of Stephen C. Schuster and Webb Miller, announced in 2008 that it had successfully sequenced the largest amount of nuclear DNA from an extinct species to date, approximately 70 percent of the woolly mammoth genome (an estimated 4.7 billion nucleotides). Though other scientists weighed in through the media to applaud the legitimate scientific milestones achieved by the Mammoth Genome Project, some felt it necessary to note that woolly mammoth cloning remained virtually impossible, largely due to the lack of an intact cell and science's inability to synthesize chromosomes in the lab. Dr. Michael Bunce, Head of the Ancient DNA Laboratory at Murdoch University, for example, noted that "just because we know the DNA code of something does not mean we can genetically tinker with it to the extent required to recreate extinct organisms—this kind of progress is still a pipe-dream."[28]

Throughout this timeframe, the prospect of mammoth cloning kept sliding back and forth across the divide between science fiction and science and it could be difficult to determine how much of the ambiguity was being driven by the mass media alone or by the mass media and some scientists in concert. In 2008, Hofreiter argued in *Nature* that "the mammoth will never roam the earth again. The best we can do is trying to understand its biology, evolution and maybe the reasons for its extinction. Perhaps this information will provide us with better tools necessary to prevent the extinction of the millions of species that currently live on our planet."[29] Yet in 2009, *National Geographic* published an article on recent woolly mammoth genome research entitled "Recipe for a Resurrection," that included an illustration of three prospective methods of bringing back a woolly mammoth: in vitro fertilization of a female Asian elephant with mammoth sperm, somatic cloning from a frozen cell, or cloning from the fully sequenced mammoth genome. Though the text of the article did note that significant scientific progress would be needed actually to accomplish this feat by any of the three proposed methods, the impact of the title, graphic, and conclusion left readers in anticipation—and the mammoth in a metaphorical state of suspended animation—pending further scientific progress that was sure to come.[30] Even more persuasively, perhaps especially for the non-expert observer, a journal article published by an international and multi-disciplinary scientific team of specialists entitled "Biological Time Machines," noted that while the probability of bringing back an extinct animal remained remote, "we must also remember that we are at the dawn of the artificial life era, or synthetic biology ... and many of the breakthroughs being achieved might complement the existing embryological procedures, thus enhancing our chances of accomplishing such a project."[31] The notion of bringing back a woolly mammoth through biotechnology had, by the end of the first decade of the twenty-first century, moved into a space somewhere between fantasy and reality. While the woolly mammoth was not with us yet, and apparently would not be until some unspecified and hazy future date, several credible scientific outlets were no longer ruling the idea completely out of bounds. In this sense, woolly mammoth ancient DNA fragments became what sociologists of science refer to as an actant, a biological entity like "immortal stem lines, genes, species, viral vectors, GMOs" that through the interventions of science now "have a certain future-orientation though of course not in quite the same cognitive sense as human expectations."[32]

The notion of bringing back a woolly mammoth, along with several other charismatic species, reached its apotheosis in March 2013, when the National Geographic Foundation in Washington, DC, co-sponsored a high-profile TEDx De-Extinction event that featured speakers such as Stewart Brand (co-executive, with his wife Ryan Phelan, of the Revive & Restore Foundation), Professor Mike Archer (Dean of Science at the University of Sydney), and paleontologist Hendrik Poinar (McMaster University) discussing progress toward the goal of de-extinction. Given the happy confluence of Silicon Valley energy, money, media visibility, and scientific prestige involved in this event, it launched de-extinction into a global arena and extended the reach of the idea far beyond that previously achieved by the thylacine cloning project. Furthermore, by including some critics of de-extinction and also bio-ethicists, the event sought to place Revive & Restore at the center of de-extinction research internationally and to provide the template for the ensuing media coverage and debates. Glossy images of the woolly mammoth on magazine covers and websites conveyed the TEDx event visually and underwrote much of the subsequent media coverage of de-extinction research and purported milestones. For example, *National Geographic*'s April 2013 issue, timed to coincide with the TEDx event, featured a stunning color image of a large woolly mammoth, followed by a cave bear, a saber-tooth tiger, a moa, a thylacine, and a dodo, all emerging from a gigantic beaker, under the title "Reviving extinct species: we can, but should we?" Evolutionary biologist Beth Shapiro, a TEDx De-Extinction speaker and expert on ancient DNA molecular analysis, published *How to Clone a Woolly Mammoth* in 2015, while the Revive & Restore Foundation showcased a Woolly Mammoth Revival project, conducted in cooperation with the lab of Professor George Church (Harvard University and MIT), on its website. *The Mammoths and Mastodons: Titans of the Ice Age* traveling exhibit, which was on the road in 2013 and included the Lyuba specimen, contained a major section that focused on the question: should we bring back the woolly mammoth? The *Washington Post* captured the prevailing spirit of both promotion and excitement in its article on the TEDx event entitled "De-extinction is the mind-blowing idea of the year" (March 21, 2013), noting that it could "open a door to breathtaking advances in genetic engineering [and] help to undo hundreds of years of ecological damage that the human species has inflicted on Planet Earth."[33]

The sophisticated multimedia use of images that indelibly linked the woolly mammoth to de-extinction constituted a new variation on the same impulse to harness the species' iconic power that previously characterized both nineteenth-century paleoimagery and the late twentieth-century innovation of "factual television program in which computer generated imagery is used extensively to bring extinct animals back to life."[34] It also put de-extinction squarely within the ecomodernist framework in that a resurrected woolly mammoth would be, simultaneously, a living, breathing example of our ingenuity in mastering the intricacies of life and death, and a sentinel of a new world in which previous environmental losses and errors could be rectified through the judicious use of biotechnologies. In an early stock take of the Anthropocene concept,

Crutzen and Stoermer noted that "in the course of history human beings have 'invited' a growing number of species to join them on their journey to a new and increasingly man-made environment, a man-made future, indeed—even into a new geological era."[35] De-extinction as an idea enlarges both the popular and scientific imagination such that extinct species, perhaps in new but recognizable hybrid forms, can also be invited into future nature. In the emerging world of life-on-demand, human aspirations are not necessarily limited to what is available today, but can reach into the past and bring genomic information and species back into a replenished wilderness. For some, these possibilities represent the essence of the promissory twenty-first biocybernetic century. As one organization, Next Nature, exults, "we must no longer see ourselves as the anti-natural species that merely threatens and eliminates nature, but rather as catalysts of evolution. With our urge to design our environment we create a 'next nature' which is unpredictable as ever: wild software, genetic surprises, autonomous machinery and splendidly beautiful black flowers. Nature changes along with us."[36] Yet, even presuming for the moment that this is both feasible and desirable, we would still be confronted, if our mammoth dreams are realized, with the anachronism of a Pleistocene species plunked down into an environment so modified from what it originally knew as to arguably constitute a different planet. It is in this temporal disjunction, as well, that the promissory woolly mammoth seems as emblematic of our time as it does of its own.

There's No Time Here, Not Any More

For a species that has been extinct for such a long time, woolly mammoths are still doing quite a lot of heavy lifting in scaffolding the de-extinction imaginary. Fletcher notes that the iconic mammoth is imminently "adaptable … both the Titan of the Ice Age, as a recent I-Max documentary puts it, and also the new symbol of an environment rapidly reaching tipping points beyond which human beings and other species may not be able to adjust."[37] This animal thus carries much of the weight of our expectations about biotechnology. Putting human beings in the frame with dinosaurs takes quite a lot of literary and cinematic effort. The woolly mammoth, though, lingers right at that crossroad between a deep time in which early humans were present, the problematic Anthropocene, and the promissory future. Moreover, as the idea of de-extinction starts to move out of its early disruptive period, dating roughly from the publication of the first quagga sequences to the TEDx De-Extinction event (1984–2013), into an era of normalization, some stakeholders now seek to de-emphasize the concept's purported revolutionary potential and forge connections instead with established conservation practice. Ben Novak, for example, one of the 2013 TEDx De-Extinction speakers and a researcher on the Great Passenger Pigeon Comeback Project (Revive & Restore), argues that if "the goal of de-extinction is to restore vital ecological functions that sustain dynamic processes producing resilient ecosystems and increasing biodiversity and bioabundance," then it "in practice is simply a continuation of restoration ecology that began in the 1830's."[38] Yet however pragmatic such a shift might be, it is a far cry from headlines such as "Back from the dead, the growth of 'resurrection science' means that a real-life Jurassic Park is no longer the stuff of science fiction" (*The Times*, March 17, 2013) that greeted the 2013 TEDx event.

Certainly, there are aspects of the broad de-extinction project, perhaps better captured in the phrase genetic rescue, that could over time enhance the tool kit available to conservation biologists and others on the frontline of species and wilderness preservation. Yet the slippage between the fantastic and the pragmatic aspects of de-extinction may prove hard to control and proper science may have a difficult task in corralling those promissory mammoths, passenger pigeons, Tasmanian tigers, and giant moas back into line. Indeed, these ancient species are so spectacular that they can obscure the extant species all around us that seem to be advancing closer to the black hole of extinction. In his TEDx De-Extinction talk in 2013, Stewart Brand said "I think that's, you know, part of how it'll work. This is a long, slow process—One of the things I like about it, it's multi-generation. We will get woolly mammoths back."[39] As will be discussed in Chap. 5, the assertion that we will get woolly mammoths back, which is made by multiple stakeholders, is contingent on nascent technologies that may or may not prove reliable, as well as on the availability of habitat, the question of whether or not an elephant-mammoth hybrid would actually be sufficient to claim success in Pleistocene de-extinction, and how strictly one insists on holding to a deadline. The pressures of the present biodiversity and extinction crises exacerbate the ethical tensions inherent in mobilizing hope in the form of a reanimated woolly mammoth but also make the notion of one almost irresistible. Extinction as an issue is too big and too urgent. It threatens to overwhelm and immobilize us. If the idea of de-extinction incentivizes new funding sources, new actors, and a renewed sense of forward momentum in addressing the extinction crisis, that would arguably be more consequential than even the arrival of a hybrid mammophant itself, and in that sense it is a difficult idea to let go of once it is out of the box. Sydney Benner argues that "by the end of the current century, we can expect that the divisions between branches of biology (molecular, cell, systems, organismic, environmental, geo- and astrobiology) will be subsumed within a broad model of the phenomenon we call life."[40] A resurrected woolly mammoth (even a hybrid) grazing in Pleistocene Park would incarnate this consolidation of bio-knowledge. It would stand as a testament to the now endless circulation of DNA through space and time and to how humans finally managed to gather up and weave together all of the bits and pieces of biology, learning not only to understand and even to modify life but also to re-ignite it and set it on a pathway of our choosing.

In *Ghosts of My Life*, Mark Fisher revisits an old British science fiction television show that aired in the 1970s, *Sapphire and Steel*. He notes that the special power of the show derived from its prescient sense of how, in the modern era, "time just got mixed, jumbled up, together, making no sort of sense. Anachronism, the slippage of discrete time periods into one another, was throughout the series the major symptom of time breaking down."[41] In this, the show was a harbinger of our own biocybernetic era, wherein digital and biotechnologies sort and recombine past, present, and future into one infinite and malleable clay. To return to Lyuba, she is such an effective representative of her species and her epoch precisely because her story can jump between times and connect our strange fascination with the Ice Age to our current trepidation in the Carbon Age. She also integrates the multiple narratives of scientists, journalists, publicists, curators, and the general public into one coherent story that is extraordinarily complex but also readily understood. Alan Mikhail argues that "various castings of the Anthropocene subsume under a single umbrella capitalism, Enlightenment, industrialization, global citizenship, the age of revolutions, the climate crisis, and the emergence of the nation-state. This is quite a lot of analytical weight for a single concept to bear."[42] I would add the extinction crisis to his list. This is also a lot of weight for Lyuba and her species to bear, but they continue to do so, remarkably.

NOTES

1. Dizikes, "Did Humans Speak Through Cave Art: Ancient Drawings and Language's Origins." See also, Miyagawa, Lesure, Nóbrega, "Cross-Modality Information Transfer: A Hypothesis About the Relationship Among Prehistoric Cave Paintings, Symbolic Thinking, and the Emergence of Language."
2. Higgins, "Ice Age Art at the British Museum Was Crafted by 'Professional' Artists."
3. Keller, "Rethinking the Meaning of Biological Information," 160.
4. DeLord, "The Nature of Extinction," 659.
5. Gee, "Memories of Mammoths," 673.
6. Holmberg, Schwennesen, Webster, "Bio-Objects and the Bio-Objectification Process," 740.
7. Dietl, Flessa, "Conservation Paleobiology: Putting the Dead to Work," 30.
8. Thacker, "The Science Fiction of Technoscience: The Politics of Simulation and a Challenge for New Media Art," 15.
9. Lubar, *Inside the Lost Museum: Curating, Past and Present*, 7–8.
10. Cuvier (1800). *Mémoires sur les espèces d'éléphants vivants et fossiles.*
11. Encyclopedia Britannica, "Fossils."
12. Coniff, "Mammoths and Mastadons: All American Monsters."
13. Steffen, Grinevald, Crutzen, McNeill, "The Anthropocene: Conceptual and Historical Perspectives."
14. Bergson, *Creative Evolution*, 153.
15. Carrington, "Earth's Sixth Mass Extinction Event Underway, Scientists Warn."
16. For a scientific analysis of the specimen Papageorgopoulou, Link, Rühli, "Histology of a Woolly Mammoth (Mammuthus primigenius) Preserved in Permafrost, Yamal Penisula, Northwest Siberia."
17. Star, Griesemer, "Institutional Ecology, 'Translations' and Boundary Objects: Amateurs and Professionals in Berkeley's Museum of Vertebrate Ecology, 1907–1939," 393.
18. National Geographic, "Waking the Baby Mammoth Facts."
19. Fletcher, "In the Hall of Extinct Monsters," 91.
20. Associated Press, "Frozen mammoth Could Shed Light on Global Warming."
21. Ibid.
22. Lowry, "Waking and Cloning: Woolly Mammoths."
23. Kolata, "So You're Extinct: Scientists Have a Gleam in the Eye."
24. Greenwood, Capelli, Possnert, Pääbo, "Nuclear DNA Sequences from Pleistocene Megafauna."

25. Greenwood, "Mammoth Biology: Biomolecules, Phylogeny, Numts, Nuclear DNA, and the Biology of an Extinct Species," 255.
26. Krause, Dear, Pollack, Slatkin, Spriggs, Barnes, et al., "Multiplex Amplification of the Mammoth Mitochondrial Genome and the Evolution of Elephantidae."
27. Poinar, Schwarz, Qi, Shapiro, MacPhee, Buiges, et al., "Metagenomics to Paleogenomics; Large-Scale Sequencing of Mammoth DNA."
28. Agence-France Press, "Mammoth Genome Cracked: Key to Cloning."
29. Hofreiter, "Mammoth Genomics," 330.
30. Mueller, "Recipe for a Resurrection."
31. Loi, Wakayama, Saragustry, Fulka, Ptak, "Biological Time Machines: A Realistic Approach for Cloning an Extinct Mammal," 233.
32. Brown, "Hope against Hype: Accountability in Biopasts, Presents and Futures," 10.
33. Basulto, "De-extinction Is the Mind-Blowing Idea of the Year."
34. Campbell, "The Extinct Animal Show: The Paleoimagery Tradition and Computer Generated Imagery in Factual Television Programs," 199.
35. Crutzen, Stoermer, "The Anthropocene," 17.
36. http://www.nextnature.net/philosophy.
37. Fletcher, "Bio-imaginaries: Bringing Back the Woolly Mammoth," 91.
38. Novak, "De-Extinction," 5.
39. Brand, "The Dawn of De-extinction: Are You Ready."
40. Benner, "Early Days of Paleogenetics: Connecting Molecules to the Planet."
41. Fisher, "The Slow Cancellation of the Future," 5.
42. Mikhail, "The Enlightenment Anthropocene," 212.

CHAPTER 5

Conclusion: Life-on-Demand

Abstract The conclusion to this book begins with a brief journey through Andy Warhol's Rain Room, an influential installation of the late 1960s that explicitly sought to integrate new technological possibilities with natural processes. As an early artistic forerunner to our age of artificial and biocybernetic life, Rain Room embodies both the potential of technology to reinvigorate our relationship to the natural world and also the more dystopian possibility of human alienation from the environment. The analysis then concludes with reflections upon the powerful allure of those promissory woolly mammoths and the question of whether our nascent era of life-on-demand is one of hope or despair.

Keywords Andy Warhol • Artificial life • Biocybernetics
• De-extinction • Mammoths • Rain Room

In 1969, Andy Warhol debuted a work entitled *Rain Machine (Daisy Waterfall)*, an installation produced for the innovative Art and Technology program at the Los Angeles County Museum of Art (LACMA). The piece, which subsequently appeared at the Tokyo Expo (1970) and roughly coincided with the first Earth Day (April 22, 1970), blends Warhol's acute appreciation for new technologies with what might be a supportive nod to the burgeoning environmental movement. In retrospect, *Rain Machine*

© The Author(s) 2020
A. L. Fletcher, *De-Extinction and the Genomics Revolution*,
https://doi.org/10.1007/978-3-030-25789-7_5

also seems to anticipate the increasingly artificial environments of our era, in which even the weather can be geo-engineered and extinct species brought back to life.

Rain Machine proved to be one of the most successful artistic-corporate collaborations of the LACMA program, in this case between Andy Warhol and Cowles Communications, facilitated by the curatorial guidance of Maurice Tuchman. Over the course of the project, four *Rain Machines* were produced. The first featured a large symmetrical array of panels, four daisies per panel, mounted on a wall behind a cascade of running water that sprayed from nozzles attached to the ceiling. In the subsequent collaboration with technicians at Cowles Communication, Warhol constructed a second *Rain Machine* with only one daisy per 3-D lenticular panel, the panels arranged into a large wall display behind continuous falling water. Both of these works disintegrated within six months of being exhibited, due to water damage. A third work that Tuchman purchased from Warhol and later donated to the Warhol Museum in Pittsburgh in the mid-1990s was irretrievably damaged while in storage. Only one original *Rain Machine* exists today. Warhol, the artist who claimed to want to be a machine, presciently created an early work of next nature that re-sensitizes the viewer to the beauty and fragility of the real thing. It also endures, albeit in digital form. In 2016, the Young Projects Gallery, located in the Pacific Design Center in West Hollywood, California, revived Andy Warhol's *Rain Machine*, the first time the work had appeared in Los Angeles in 45 years. In this case, the contemporary artist Refik Anadol reimagined *Rain Machine* as a six-room immersive installation in which the viewer walks through digitized rain and storms, surrounded by ambient noise that mixes natural soundscapes and classical music. In the last room, the viewer finds Warhol's original *Rain Machine*, complete with the front-facing trough that in earlier iterations of the piece caught real water as it cascaded down the lenticular daisy panel.

Anadol noted that "collaborating, posthumously, with Warhol, 'is like a dream.' It's super meaningful. He's one of those artists who [shapes] people's minds and changes perceptions of contemporary art."[1] This observation connects *Rain Machine* firmly to this contemporary era of both artificial life and the Anthropocene and suggests important parallels with the de-extinction project. Whether revitalizing analog artworks or attempting to resuscitate extinct species, many scientists and artists today seem entranced with re-constituting the past as a prelude to moving into a more sustainable future. Warhol seems to have intuited earlier than most that not only human relationships and identities would be increasingly mediated by technology in the twenty-first century, but that Nature itself could be conceptualized as a machine, subject to virtually endless human engineering. *Rain Machine* is emblematic of a world moving toward life-on-demand, though whether Warhol, who once purportedly said "I don't know where the artificial stops and the real starts," would have celebrated or resisted this development is unclear. The piece also appeared in the 1990 exhibit *Artificial Nature*, curated by Jeffrey Deitch and exhibited at the Deste Foundation in Athens, Greece. The catalog for that exhibit asserts that "with the direction of science toward the creation of artificial life forms and a computerized virtual reality ... there is no longer one absolute reality. There is only the possibility of multiple realities, each one as 'real' or as artificial as the others."[2]

The implications of this shift are profound, but we have been anticipating this outcome for quite some time. Heinz Heck and Andy Warhol, representatives of science and spectacle, respectively, anticipate human control of nature in the middle of the twentieth century, building on dreams that began in the eighteenth century. That future now seems to be arriving, on an unreliable schedule perhaps, but always moving forward into next nature. Indeed, Random International, a digital art group based in London and Berlin, now offers the art market a Rain Room that "can be seen as an amplified representation of our environment. Human presence prevents the rain from falling, creating a unique atmosphere and exploring how human relationships to each other and to nature are increasingly mediated through technology."[3] In 2016, an unauthorized replica of Rain Room was installed in the Jinjiale Dream Park in Shanghai, though the question of who owns the digital rain had yet to be adjudicated.

This book concludes with some brief reflections upon the emerging era of life-on-demand and the powerful pull of those promissory woolly mammoths who, like Rain Room and other emergent forms of techno-life, exist somewhere between the real and the artificial, in that space wherein the future is always both imminent and malleable. By approximately 2015, many journalists began to be more judicious regarding reports of woolly mammoth cloning. Certainly, Beth Shapiro's *How to Clone a Woolly Mammoth*, which actually details the reasons that the pathway from ancient DNA fragments to an intact cell makes cloning such an ancient beast all but impossible, was instrumental in pulling paleogenomics back from the brink of hype into the realms of professional science. However, the engineering mindset brings with it a fundamental optimism about finally pushing past barriers: if nature is indeed a machine, then it should be able to be reverse-engineered with enough ingenuity and funding and some patience. Though the type of reproductive cloning that produced Dolly the Sheep seems an increasingly remote possibility for bringing back a woolly mammoth, synthetic biology and the powerful gene-editing tool CRISPR-Cas 9 combine to produce an alternative pathway for resurrecting a mammoth-elephant hybrid that, to most observers, would be indistinguishable from the Pleistocene original.

THE GHOSTS OF MAMMOTHS YET TO COME

As the northern white rhino fades into oblivion on our watch, collecting and freezing skin samples of the species in the hopes that reproductive biotechnologies could provide a solution in the future is an urgent and necessary task, a wager against despair. As Philip Seddon argues:

> to do better—even if that means nothing more than failing less spectacularly—bolder thinking is necessary. Among the strategies we should consider is engagement with the rapidly expanding field of synthetic biology, whose genetic editing tools suggest new techniques for protecting threatened species and limiting invasive one, addressing what are currently intractable challenges in conservation.[4]

The broad field of genetic rescue—and de-extinction—does encompass a scientific search for tools to fight the extinction crisis that advances through the incremental findings of credentialed expertise, field trials, and refereed journal articles. However, those dreams of woolly mammoths yet to come are quite powerful and this collision between hope and hype and science and spectacle continues to generate sparks in the science media. The development of CRISPR-Cas 9, for example, which "allows the 'cut and paste' manipulation of strands of DNA with a precision not seen before,"[5] enables research into splicing the genome of an Asian elephant with genetic information sourced in the lab from the woolly mammoth sequence. In 2015, in a provocative turn of phrase, *Digital Journal* reported that "for the first time in over 3,000 years, mammoth DNA is alive and the aim of cloning the ancient animal is now much closer to reality."[6] The article discussed research from George Church's lab which used CRISPR to insert 14 woolly mammoth genes into the genome of an Asian elephant, including genes that coded for thick woolly coats, subcutaneous fat, ear size, and other traits that appear to have built mammoths for the

Siberian climate. Phys.org was more cautious in its appraisal of the work, noting a mammoth was not forthcoming at this stage. Still, it noted that "what they have done, however, is build healthy living cells with mammoth DNA in them. Their work is yet another step towards that ultimate goal, realizing the birth of a woolly mammoth that is as faithful to the original as is humanly possible."[7] By 2018, the Church Lab had "resurrected" 44 mammoth genes, with George Church arguing that "my goal is not to bring back the mammoth, it's to bring back mammoth genes and show that they work … We want something that can adapt to a different environment so we save two ecosystems, one is the elephant ecosystem and the other is the tundra"[8] (Image 5.1).

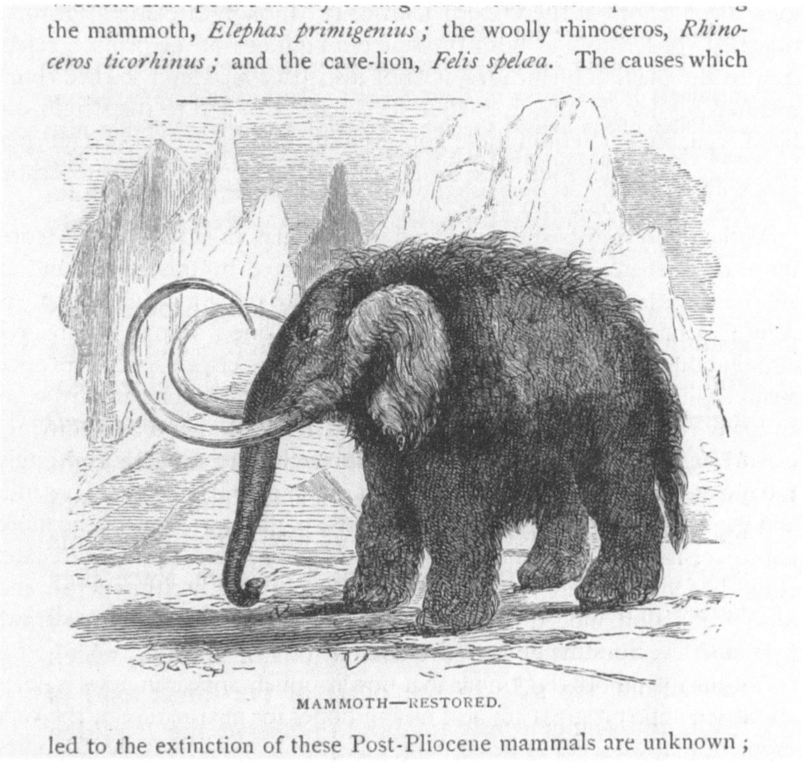

the mammoth, *Elephas primigenius;* the woolly rhinoceros, *Rhinoceros ticorhinus;* and the cave-lion, *Felis spelæa.* The causes which

MAMMOTH—RESTORED.

led to the extinction of these Post-Pliocene mammals are unknown ;

Image 5.1 Mammoth restored, from *Beneath the Surface: or the Wonders of the Underground World*, 1876. Author William Henry Davenport (Source: British Library. Public Domain)

From one perspective, the new emphasis on gene editing, as opposed to reproductive cloning, moved woolly mammoth resurrection more firmly into the sphere of genetic rescue and deep ecological enrichment and away from the more fantastical claims of de-extinction. However, not everyone was convinced that a revivified mammophant, if it happened at all, would significantly improve either the prospects for the critically endangered Asian elephant or make a demonstrable difference to the permafrost in Siberia and, by extension, climate emissions mitigation. Tori Herridge, a paleobiologist and mammoth expert, cautions that "it is a big gamble to put your climate-change mitigation hopes on a herd of woolly mammoths—and if it did work, it would require numbers in the hundreds of thousands to have an effect."[9] Mark Carnall, Collections Manager of zoological collections at the Oxford University Museum of Natural History, objected vociferously to both the stated aspirations of the project, referring to the mammophant idea as "not just space race level science [but] manned Jupiter-colonizing mission level science"[10] and to the sleight-of-hand arguably inherent in these notions of hybridized extinct/extant species, suggesting "let's just tape some carpet to an elephant and photoshop it into a tundra scene and move on."[11]

While the scientists and associated professionals in natural history continue this debate, most of us who are captivated by this issue stand at almost the exact same observation point where our forebears began. As Amy Fletcher argues, we are "in the admission line to future nature, yet also find ourselves back at P.T. Barnum's three-ring circus, spectators once again trying to parse the line between fact and fantasy, hope and hype."[12] In many respects, following this particular story is fun and also edifying. Cavorting with woolly mammoths and Tasmanian tigers appeals as strongly today as it did when Peale opened his first museum in Baltimore, Maryland, and whether in the more approachable guise of genetic rescue or the more controversial de-extinction, the science being done here, in the aggregate, could benefit not only conservation but also human health, energy, and food production, indeed virtually any sphere of human activity that draws from and leverages the power of nature.

Yet the mammoths and moas that now routinely appear in news reports about scientific progress are also raising hopes for next nature at the very moment that we seem to be running out of time. Shelley, Peale, Barnum, and their ilk did not enjoy easier times than ours, indeed in many practical ways they were much harder and more dangerous; however, they did enjoy more expansive times, particularly where science and its interactions with society were concerned. Whether in the serious science of Charles Darwin or the shenanigans of P.T. Barnum, the spirit of inquiry and a sense of wonder bubble through their writings. The twentieth century was not so lucky in this regard. By 1920, Sir Ray E. Lankester, one of those learned amateur naturalists that flourished in this era, warned that "very few people have any idea of the extent to which man … has actively modified the face of Nature, the vast herds of animals he has destroyed, the forests he has burnt up, the deserts he has produced, and the rivers he has polluted."[13] Interestingly, Lankester adds that "no compensatory production of new life, except that of man himself and his distorted 'breeds' of domesticated animals, has accompanied the destruction of formerly flourishing creatures."[14] In certain ways, de-extinction and genetic rescue take up the call to produce that new (or renewed) life and to replenish the Earth with the charismatic species whose legends persist through the centuries. For this reason, it is imperative that we approach the hype with the appropriate skepticism but also let de-extinction renew our sense of wonder and hope. I want to give the last word to Japanese scientist Akira Iritani, age 90, who in retirement had almost given up on his desire to push the scientific frontiers of potential woolly mammoth resurrection but whose curiosity was spurred by the discovery in 2010 of a 28,000-year-old specimen named Yuka. In collaboration with Russian scientists, his team used live-cell imaging to see if long-dormant samples collected from the mammoth might react when interjected into mouse oocytes. Like all such research, the findings are provisional and need to be subject to replication, but the cells did apparently stir into life in the lab. In the words of Professor Iritani, "I'm so happy with this latest research. It feels like Yuka was waiting for me to find her."[15]

Notes

1. Vankin, "Step Inside a Digital Storm."
2. Dietch, "Artificial Nature."
3. https://www.random-international.com/rain-room-2012.
4. Seddon, "De-extinction and Barriers to the Application of New Conservation Tools," S5.
5. Devlin, "Woolly on Verge of Resurrection, Scientists Reveal."
6. Morgan, "Woolly Mammoth Rebirth Much Nearer, Cells Alive in the Laboratory."
7. Yirka, "Researchers Take Another Step in Bringing Back a Woolly Mammoth."

8. Knapton, "Scientists on the Verge of Creating Hybrid Elephant and Mammoth."
9. Herridge, "Mammoths Are a Huge Part of My Life. And Cloning Them Is Wrong."
10. Carnall, "Undoing Extinction—Let's Talk about the Mammophant in the Room."
11. Ibid.
12. Fletcher, "In the Hall of Extinct Monsters," 143.
13. Lankester, *More Science from an Easy Chair.*
14. Ibid.
15. Jozuka, "The 90-Year Old Japanese Scientist Still Dreaming of Resurrecting a Woolly Mammoth."

BIBLIOGRAPHY

Agence-France Press. 2002. Bid to Clone Extinct 'Tasmanian Tiger' Assailed. *Agence-France Press*, August 22.

———. 2008. Mammoth Genome Cracked: Key to Cloning. *Cosmos*, 2018. http://archive.cosmosmagazine.com/news/mammoth-genome-cracked-key-cloning. Accessed April 17, 2016.

Asafu-Adjaye, John, Linus Blomqvist, Stewart Brand, Barry Brook, Ruth DeFries, and Erle Ellis. 2015. *The Ecomodernist Manifesto.* http://www.ecomodernism.org/. Accessed April 30, 2019.

Associated Press. 2008. Frozen Mammoth Could Shed Light on Global Warming. *CTV News*, January 4. https://www.ctvnews.ca/frozen-mammoth-could-shed-light-on-global-warming-1.269917. Accessed May 31, 2018.

Barnum, P.T. 1866. *The Humbugs of the World: An Account of Humbugs, Delusions, Impositions, Quackeries, Deceits and Deceivers Generally, in All Ages.* https://en.wikisource.org/wiki/The_Humbugs_of_the_World. Accessed April 30, 2019.

Baron, Zach. 2016. Inside the Frozen Zoo That Could Bring Extinct Animals Back to Life. *GQ*, October 27. https://www.gq.com/story/inside-the-frozen-zoo-that-could-bring-extinct-animals-back-to-life. Accessed April 30, 2019.

Basulto, Dominic. 2013. De-Extinction Is the Mind-Blowing Idea of the Year. *The Washington Post*, March 21. https://www.washingtonpost.com/blogs/innovations/post/de-extinction-is-the-mind-blowing-idea-of-the-year/2013/03/19/cbbce3b8-908e-11e2-9173-7f87cda73b49_blog.html?noredirect=on&utm_term=.2863452c5894. Accessed April 26, 2019.

Baudrillard, Jean. 1983. *Simulations.* Cambridge: The MIT Press.

© The Author(s) 2020

A. L. Fletcher, *De-Extinction and the Genomics Revolution*, https://doi.org/10.1007/978-3-030-25789-7

Bearak, Max. 2018. Sudan, the World's Last Male Northern White Rhino, Has Died, Putting His Species on the Brink of Extinction. *The Washington Post*, March 20. https://www.washingtonpost.com/news/worldviews/wp/2018/03/20/sudan-the-worlds-last-male-northern-white-rhino-has-died-putting-his-species-on-the-brink-of-extinction/?noredirect=on&utm_term=.eac57a5574cb. Accessed April 22, 2019.

Benner, Steven. 2007. The Early Days of Paleogenetics: Connecting Molecules to the Planet. In *Ancestral Sequence Reconstruction*, ed. David A. Liberles. Oxford: Oxford University Press. https://doi.org/10.1093/acprof:oso/9780199299188.003.0001.

Bergson, Henri. 1911. *Creative Evolution*. Trans. A. Mitchell. New York: The Modern Library. Reprinted in 1944. Original *L'Evolution Créatrice*, 1907.

Boissoneault, Lorraine. 2018. Jurassic Park's Unlikely Symbiosis with Real-World Science. *Smithsonian.com*, June 15. https://www.smithsonianmag.com/science-nature/jurassic-park-reveals-delicate-interplay-between-science-and-science-fiction-180969331/. Accessed April 22, 2019.

Bradshaw, Corey, and J. A. De. 2013. Extinction Is About as Sensible as De-Death. *The Conversation*, March 15. https://theconversation.com/de-extinction-is-about-as-sensible-as-de-death-12850. Accessed April 29, 2019.

Brand, Stewart. 2013. The Dawn of De-Extinction: Are You Ready. *TEDx*, March 15. https://www.ted.com/talks/stewart_brand_the_dawn_of_de_extinction_are_you_ready/transcript?language=en. Accessed April 26, 2019.

Brenner, Sydney. 2001. *My Life in Science*. Dordrecht: Biomed Central Limited.

Briggs, Helen. 2005. Extinct Cave Bear DNA Sequenced. *The Guardian*, June 3. http://news.bbc.co.uk/2/hi/science/nature/4602739.stm. Accessed April 30, 2019.

Brough, John Cargill. 1857. *The Fairy Tales of Science: The Age of Monsters*. London: Griffin and Farran. https://en.wikisource.org/wiki/The_fairy_tales_of_science. Accessed April 28, 2019.

Brown, Nik. 2003. Hope Against Hype: Accountability in Biopasts, Presents and Futures. *Science Studies* 16 (3): 3–21.

C. elegans Sequencing Consortium. 1998. Genome Sequencing of the Nematode *C. elegans*: A Platform for Investigating Biology. *Science* 282 (5396): 2012–2018.

Campbell, Vincent. 2008. The Extinct Animal Show: The Paleoimagery Tradition and Computer-Generated Imagery in Factual Television Programs. *Public Understanding of Science* 18 (2): 199–213. https://doi.org/10.1177/0963662507081246.

Cano, Raul J., Hendrik N. Poinar, Norman J. Pieniazek, Aftim Acra, and George O. Poinar Jr. 1993. Amplification and Sequencing of DNA from a 120-135-Million-Year-Old Weevil. *Nature* 363: 536–538.

Carnal, Mark. 2017. Undoing Extinction—Let's Talk About the Mammophant in the Room. *The Guardian*, February 22. https://www.theguardian.com/science/2017/feb/22/undoing-extinction-mammoth-dextinction. Accessed April 29, 2019.

Carrington, Damien. 2017. Earth's Sixth Mass Extinction Event Underway, Scientists Warn. *The Guardian*, July 10. https://www.theguardian.com/environment/2017/jul/10/earths-sixth-mass-extinction-event-already-under-way-scientists-warn. Accessed May 31, 2018.

Cha, Ariana Eunjung. 2015. Peter Thiel's Quest to Find the Key to Eternal Life. *The Washington Post*, April 3. https://www.washingtonpost.com/business/on-leadership/peter-thiels-life-goal-to-extend-our-time-on-this-earth/2015/04/03/b7a1779c-4814-11e4-891d-713f052086a0_story.html?noredirect=on&utm_term=.89e0d14696be. Accessed April 29, 2019.

Chemical Engineering News. 1984. Gene Fragments from Extinct Animal Cloned. *C&EN Archives*. https://pubs.acs.org/doi/abs/10.1021/cen-v062n024.p008a. Accessed April 23, 2019.

Cobb, Matthew. 2013. 1953: When Genes Became Information. *Cell* 153 (April 25): 503–506.

Conniff, Richard. 2010. Mammoths and Mastodons: All American Monsters. *Smithsonian.com*, April. https://www.smithsonianmag.com/science-nature/mammoths-and-mastodons-all-american-monsters-8898672/. Accessed March 20, 2019.

Crutzen, P.J., and E.F. Stoermer. 2000. The Anthropocene. *Global Change Newsletter* 41: 17.

Cuvier, Georges. 1800. Mémoires sur les espèces d'éléphants vivants et fossiles. https://www.biodiversitylibrary.org/page/16303001#page/175/mode/1up. Accessed April 29, 2018.

DeBlonde, Martin, et al. 2008. Co-creating Nano-Imaginaries: Reports of a Delphi-Exercise. *Bulletin of Science, Technology & Society* 28 (5): 372–389.

Deitch, Jeffrey. 1990. Artificial Nature. http://www.deitch.com/curatorial/artificial-nature. Accessed September 24, 2018.

Delord, Jacques. 2007. The Nature of Extinction. *Studies in the History and Philosophy of Biomedical Sciences* 38 (3): 656–667.

Devlin, Hannah. 2017. Woolly Mammoth on Verge of Resurrection, Scientists Reveal. *The Guardian*, February 16. https://www.theguardian.com/science/2017/feb/16/woolly-mammoth-resurrection-scientists. Accessed April 22, 2019.

Diechmann, Ute. 2010. Gemmules and Elements: On Darwin's and Mendel's Concepts and Methods in Heredity. In *Darwinism, Philosophy, and Experimental Biology*, ed. Ute Diechmann and Anthony S. Travis, 31–58. Dordrecht: Springer.

Dietl, G.P., and K. Flessa. 2011. Conservation Paleobiology: Putting the Dead to Work. *Geosciences* 26 (1): 30–37. https://doi.org/10.1016/j.tree.2010.09.010.

Dizikes, Peter. 2018. Did Humans Speak Through Cave Art: Ancient Drawings and Language's Origins. *Science Daily*, February 21. https://www.science-daily.com/releases/2018/02/180221122923.htm. Accessed April 24, 2019.

Donlan, C.J., J. Berger, C.E. Bock, J.H. Bock, D.A. Berney, and J.A. Estes. 2006. Pleistocene Rewilding: An Optimistic Agenda for Twenty-First Century Conservation. *The American Naturalist* 168 (5): 660–661.

Doward, Jamie. 2018. Who Put the Spark in Frankenstein's Monster? *The Guardian*, March 4. https://www.theguardian.com/books/2018/mar/04/frankenstein-monster-200th-anniversary-electricity-mary-shelley. Accessed April 22, 2019.

Dryzek, John S. 1997. *The Politics of the Earth: Environmental Discourses.* Oxford: Oxford University Press.

Encyclopedia Americana. 1920. Fossils. https://en.wikisource.org/wiki/The_Encyclopedia_Americana_(1920)/Fossils. Accessed April 25, 2019.

Engen, Josh. 2014. Woolly Mammoth Clones: Arriving Soon. *The Escapist*, February 28. https://www.escapistmagazine.com/news/view/132558-Woolly-Mammoth-Clones-Arriving-Soon. Accessed April 29, 2019.

Evans, R., I. Kotchetkova, and S. Langer. 2008. Just Around the Corner: Rhetorics of Progress and Promise in Genetic Research. *Public Understanding of Science* 18 (1): 43–59.

Fischer, Mark. 2014. The Slow Cancellation of the Future. In *Ghosts of My Life: Writings on Depression, Hauntology, and Lost Futures*, 2–29. London: John Hunt Publishing.

Fletcher, Amy Lynn. 2010. Genuine Fakes: Cloning Extinct Species as Science and Spectacle. *Politics and Life Sciences Journal* 29 (1): 48–60. https://doi.org/10.2990/29_1_48.

———. 2014. Bio-Imaginaries: Bringing Back the Woolly Mammoth. In *Mendel's Ark: Biotechnology and the Future of Extinction*, 89–97. Dordrecht: Springer.

———. 2016. In the Hall of Extinct Monsters. In *Conflict, Negotiation and Co-Existence: Rethinking Human-Elephant Relations in South Asia*, ed. Piers Locke and Jane Buckingham. Oxford: Oxford University Press.

Frederick, Dan. 2014. Should Scientists Bring Back the Woolly Mammoth. *DW*, April 29. https://doi.org/10.3390/genes9110548.

Garcia-Sancho, M. 2007. Mapping and Sequencing Information: The Social Context for the Genomics Revolution. *Endeavour* 31 (1): 18–23.

Gee, Henry. 2006. Memories of Mammoths. *Nature* 439: 673. https://doi.org/10.1038/439673a.

Gieryn, Thomas. 1995. Boundaries of Science. In *Handbook of Science and Technology Studies*, ed. Sheila Jasanoff, Gerald E. Markle, James C. Peterson, and Trevor Pinch, 392–443. London: Sage Publications.

Gray, John MacLachlan. 2011. *The Immortalization Commission: Science and the Strange Quest to Cheat Death*. New York: Farrar, Straus and Giroux.

Greenwood, Alex D. 2000. Mammoth Biology: Biomolecules, Phylogeny, Numts, Nuclear DNA, and the Biology of an Extinct Species. *Ancient Biomolecules* 3: 255–266.

Greenwood, A.D., C. Capelli, G. Possnert, and S. Pääbo. 1999. Nuclear DNA Sequences from Late Pleistocene Megafauna. *Molecular Biology and Evolution* 16 (11): 1466–1473.

Gutiérrez, G., and A. Márin. 1998. The Most Ancient DNA Recovered from an Amber-Preserved Specimen May Not Be as Ancient as It Seems. *Molecular Biology & Evolution* 15 (7): 926–929. https://doi.org/10.1093/oxfordjournals.molbev.a025998.

Hall, Sean. 2018. Resurrecting the Northern White Rhino and Other Species. But at What Cost. *Genetic Literacy Project*, July 10. https://geneticliteracyproject.org/2018/07/10/resurrecting-the-northern-white-rhino-and-other-lost-species-but-at-what-cost/. Accessed April 22, 2019.

Hawks, John. 2017. How Mammoth Cloning Became Fake News. *Medium*, February 19. https://medium.com/@johnhawks/how-mammoth-cloning-became-fake-news-1e3a80e54d42. Accessed April 22, 2019.

Heck, Heinz. 1951. The Breeding-Back of the Aurochs. *Oryx* 1 (3): 117–122. Trans. Winifred Felce.

Hedgecock, Sarah. 2017. What the Woolly Mammoth De-Extinction Project Actually Means. *Forbes*, March 1. https://www.forbes.com/sites/sarahhedgecock/2017/03/01/what-the-woolly-mammoth-de-extinction-project-actually-means/#52ca89304a61. Accessed April 29, 2019.

Heinemann, Pia. 2017. Extinction Was Yesterday. *Welt*, July 8. https://www.welt.de/wissenschaft/article166329856/Extinction-was-yesterday.html. Accessed April 30, 2019.

Herridge, Tori. 2014. Mammoths Are a Huge Part of My Life. But Cloning Them Is Wrong. *The Guardian*, November 18. https://www.theguardian.com/commentisfree/2014/nov/18/mammoth-cloning-wrong-save-endangered-elephants. Accessed April 29, 2019.

Higgins, Charlotte. 2013. Ice Age Art at the British Museum Was Crafted by 'Professional' Artists. *The Guardian*, January 24. https://www.theguardian.com/science/2013/jan/24/ice-age-art-british-museum. Accessed April 25, 2019.

Higuchi, R., B. Bowman, M. Frieburger, O.A. Ryder, and A.C. Wilson. 1984. DNA Sequences from the Quagga, an Extinct Member of the Horse Family. *Nature* 312. https://doi.org/10.1038/312282a0.

Hills, A., and A. Rosenfeld. 1963. Nearer Now: Control of Aging and Heredity. *Life International* 39 (5): 45–50.

Hofreiter, M. 2008. Mammoth Genomics. *Nature* 456: 330–331.

Holmberg, Tora, Nete Schwennesen, and Andrew Webster. 2011. Bio-Objects and the Bio-Objectification Process. *Croatian Medical Journal* 52 (6): 740–742. https://doi.org/10.3325/cmj.2011.52.740.

Hood, Marlowe. 2019. Many Sharks Closer to Extinction Than Feared: Red List. *Phys.org*, March 22. https://phys.org/news/2019-03-sharks-closer-extinction-red.html. Accessed April 22, 2019.

Jiang, Kevin. 2015. The Genes That Make a Woolly Mammoth a Woolly Mammoth. *Science Life*, July 2. https://sciencelife.uchospitals.edu/2015/07/02/the-genes-that-make-a-woolly-mammoth-a-woolly-mammoth/. Accessed April 29, 2019.

Jokuza, Emiko. 2019. The 90-Year-Old Japanese Scientist Still Dreaming of Resurrecting a Woolly Mammoth. *CNN*, March 18. https://edition.cnn.com/2019/03/18/health/japan-woolly-mammoth-resurrection-intl/index.html. Accessed May 1, 2019.

Keller, Evelyn Fox. 2002. *The Century of the Gene*. Cambridge, MA: Harvard University Press.

———. 2009. Rethinking the Meaning of Biological Information. *Biological Theory* 4 (2): 159–166.

Knapton, S. 2017. Woolly Mammoth Will Be Back from Extinction in Two Years, Say Harvard Scientists. *The Telegraph*, February 17. https://www.telegraph.co.uk/science/2017/02/16/harvard-scientists-pledge-bring-back-woolly-mammoth-extinction/. Accessed April 22, 2019.

———. 2018. Scientists on the Verge of Creating Hybrid Elephant and Mammoth. *The Sydney Morning Herald*, April 29. https://www.smh.com.au/world/europe/scientists-on-the-verge-of-creating-hybrid-elephant-and-mammoth-20180429-p4zca6.html. Accessed April 29, 2019.

Kolata, Gina. 1997. Scientists Report First Cloning Ever of Adult Mammal. *The New York Times*, February 23. https://www.nytimes.com/1997/02/23/us/scientist-reports-first-cloning-ever-of-adult-mammal.html. Accessed April 23, 2019.

———. 2013. So You're Extinct: Scientists Have a Gleam in the Eye. *The New York Times*, March 18. https://www.nytimes.com/2013/03/19/science/earth/research-to-bring-back-extinct-frog-points-to-new-path-and-quandaries.html. Accessed March 10, 2019.

Krause, J., and Svante Pääbo. 2016. Genetic Time Travel. *Genetics* 203 (1): 9–12. https://doi.org/10.1534/genetics.116.187856. Accessed April 23, 2019.

Krause, J., P.H. Dear, J.L. Pollack, M. Slatkin, I. Barnes, A.M. Lister, et al. 2009. Multiplex Amplification of the Mammoth Mitochondrial Genome and the Evolution of Elephantidae. *Nature* 439 (7077): 724–727.

Lankester, Sir E. Ray. 1920. *More Science from an Easy Chair*. London: Methuen & Co., Ltd. First edition 1913. http://www.gutenberg.org/ebooks/27015. Accessed May 1, 2019.

Lichtenstein, Jesse. 2012. The Synthesizers. *Tin House* 13 (3): 147–165.

Life International. 1963. DNA's Code: Key to All Life. *Life International* 35 (9): 38–43.

Lightman, Bernard. 2007. *Victorian Popularizers of Science: Designing Nature for New Audiences*. Chicago: The University of Chicago Press.

Loi, P., T. Wakayama, J. Saragustry, J. Fulka Jr., and G. Ptak. 2011. Biological Time Machines: A Realistic Approach for Cloning an Extinct Mammal. *Endangered Species Research* 14 (3): 227–233.

Lowry, Sam. 2009. Waking and Cloning: Woolly Mammoths. *Discover*, April 22. http://blogs.discovermagazine.com/sciencenotfiction/2009/04/22/waking-and-cloning-baby-mammoths/#.Ww945J8zYow. Accessed May 31, 2018.

Lubar, Steven. 2017. *Inside the Lost Museum: Curating, Past and Present*. Cambridge: Harvard University Press.

Lynch, V.J., O.C. Bedoya-Reina, A. Ratan, M. Sulak, D.I. Drautz-Moses, G.H. Perry, et al. 2015. Elephantid Genomes Reveal the Molecular Bases of Woolly Mammoth Adaptations to the Arctic. *Cell Reports* 12 (2): 217–228.

McBeth, M.K., E.A. Shanahan, and M.D. Jones. 2005a. The Science of Storytelling: Measuring Policy Beliefs in Greater Yellowstone. *Society & Natural Resources* 18 (5): 413–429. https://doi.org/10.1080/08941920590924765.

———. 2005b. The Science of Storytelling: Measuring Policy Beliefs in Greater Yellowstone. *Society and Natural Resources* 18: 413–429.

McGregor, Andrew. 2004. Sustainable Development and 'Warm, Fuzzy Feelings': Discourse and Nature Within Australian Environmental Imaginaries. *Geoforum* 35 (5): 593–606. https://doi.org/10.1016/j.geoforum.2004.02.001.

Meyer, Michal. 2012. Immortal Trouble. *Distillations*. https://www.sciencehistory.org/distillations/magazine/immortal-trouble. Accessed April 30, 2019.

Mikhail, Alan. 2016. The Enlightenment Anthropocene. *Eighteenth Century Studies* 49 (2): 211–231.

Mitchell, W.J.T. 2006. *What Do Pictures Want: The Life and Loves of Images*. Chicago: The University of Chicago Press.

Miyaga, S., C. Lesure, and V.A. Nóbrega. 2018. Cross-Modality Information Transfer: A Hypothesis About the Relationship Among Prehistoric Cave Paintings, Symbolic Thinking, and the Emergence of Language. *Frontiers in Psychology*. https://doi.org/10.3389/fpsyg.2018.00115.

Moodley, Y., I.-R.M. Russo, J. Robovsky, D. Dalton, A. Kotzé, S. Stejskal Smith, et al. 2018. Contrasting Evolutionary History, Anthropogenic Declines and Genetic Contact in the Northern and Southern White Rhinoceros (*Ceratotherium simum*). *Proceedings of the Royal Society B: Biological Sciences* 285 (1890). https://doi.org/10.1098/rspb.2018.1567.

Morgan, Stephen. 2015. Woolly Mammoth Rebirth Much Nearer: Cells Alive in the Laboratory. *Digital Journal*, March 23. http://www.digitaljournal.com/science/woolly-mammoth-rebirth-much-nearer-cells-now-alive-in-laboratory/article/428967#ixzz5mMiVhrIF. Accessed April 28, 2019.

Mueller, Tom. 2009. Recipe for a Resurrection. *National Geographic*, May. https://www.nationalgeographic.com/magazine/2009/05/cloned-species/. Accessed April 26, 2019.

Narasimhan, S.D. 2015. Resurrection. *Cell* 162: 229–231.

National Geographic. 2013. Waking the Baby Mammoth Facts. http://channel.nationalgeographic.com/a-night-of-exploration/articles/waking-the-baby-mammoth-facts/. Accessed May 31, 2018.

Novak, Ben J. 2018. De-Extinction. *Genes* 9: 11. https://doi.org/10.3390/genes9110548.

Nowlin, C. 2018. 200 Years After Frankenstein. *Perspectives in Biology & Medicine* 61: 430–449.

Pääbo, S. 1985. Molecular Cloning of Ancient Egyptian Mummy DNA. *Nature* 314: 644–645. https://www.nature.com/articles/314644a0. Accessed April 23, 2019.

Pääbo, S., H. Poinar, D. Serre, V. Jaenicke-Despres, J. Hebler, N. Rohland, et al. 2004. Genetic Analyses from Ancient DNA. *Annual Review of Genetics* 38: 645–679. https://doi.org/10.1146/annurev.genet.37.110801.143214. Accessed April 23, 2019.

Papageorgopoulou, C., K. Link, and F.J. Ruhli. 2015. Histology of a Woolly Mammoth (*Mammuthus primigenius*) Preserved in Permafrost, Yamal Penisula, Northwest Siberia. *The Anatomical Record* 298 (6): 1059–1071.

Pelis, Kim. 1999. Transfusion with Teeth. In *Manifesting Medicine: Bodies and Machines*, ed. R. Bud, B.S. Finn, and H. Trischler, 1–30. Oxfordshire: Taylor & Francis.

Petersen, A. 2001. Biofantasies: Genetics and Medicine in the Print News Media. *Social Science & Medicine* 52 (8): 1255–1268.

Poinar, Hendrik N., Carsten Schwarz, Qi Ji, J. Beth Shapiro, Ross D.E. MacPhee, Bernard Buiges, et al. 2006. Metagenomics to Paleogenomics: Large-Scale Sequencing of Mammoth DNA. *Science* 311 (5759): 392–394. https://doi.org/10.1126/science.1123360.

Purdy, Jedediah. 2015. *After Nature: A Politics for the Anthropocene*. Cambridge: Harvard University Press.

Rabinowitz, Alan. 2010. A New Strategy for Saving the World's Wild Big Cats. *Yale Environment 360*, February 27. https://e360.yale.edu/features/a_new_strategy_for_saving_the_worlds_wild_big_cats.

Reardon, Sara. 2019, April 17. Pig Brains Kept Alive Outside Body for Hours After Death. *Nature* 568: 283–284. https://www.nature.com/articles/d41586-019-01216-4. Accessed April 29, 2019.

Regalado, Antonio. 2018. Rewriting Life: A Stealthy Human Startup Wants to Reverse Aging in Dogs, and Humans Could Be Next. *MIT Technology Review*, May 9. https://www.technologyreview.com/s/611018/a-stealthy-harvard-startup-wants-to-reverse-aging-in-dogs-and-humans-could-be-next/. April 29, 2019.

Rich, Nathaniel. 2014. The Mammoth Cometh. *The New York Times*, February 27. https://www.nytimes.com/2014/03/02/magazine/the-mammoth-cometh.html. Accessed April 29, 2019.

Rodrigues, Olivia. 2019. How Jurassic Park Has Changed the Way We Exhibit Dinosaurs. *Frieze*, April 18. http://www.frieze.info/article/how-jurassic-park-has-changed-way-we-exhibit-dinosaurs?language=de. Accessed April 29, 2019.

Rose, Nikolas. 2007. *The Politics of Life Itself: Biomedicine, Power, and Subjectivity in the Twenty-First Century*. Princeton: Princeton University Press.

Ruston, Sharon. 2014. The Science of Life and Death in Mary Shelley's *Frankenstein*. British Library: Discovering Literature: Romantics & Victorians, May 15. https://www.bl.uk/romantics-and-victorians/articles/the-science-of-life-and-death-in-mary-shelleys-frankenstein. Accessed April 22, 2019.

Sánchez-Bayo, Francisco, and Kris A.G. Wyckhuys. 2019. Worldwide Decline of the Entomofauna: A Review of Its Drivers. *Biological Conservation* 232: 8–27. https://doi.org/10.1016/j.biocon.2019.01.020.

Sarchet, Penny. 2017. Can We Grow Woolly Mammoths in the Lab: George Church Hopes So. *The New Scientist*. https://www.newscientist.com/article/2121503-can-we-grow-woolly-mammoths-in-the-lab-george-church-hopes-so/. Accessed April 28, 2019.

Save the Rhino. 2017. Can We Save the Northern White Rhino, March 1. https://www.savetherhino.org/thorny-issues/can-we-save-the-northern-white-rhino/. Accessed April 22, 2019.

Scheele, B.C., F. Pasmans, L.F. Skerratt, L. Berger, A. Martel, W. Beukema, et al. 2019. Amphibian Fungal Panzootic Causes Catastrophic and Ongoing Loss of Biodiversity. *Science* 363 (6434): 1469–1463. https://doi.org/10.1126/science.aav0379.

Schrodinger, Erwin. 1967. *What Is Life: The Physical Aspect of the Living Cell with Mind and Matter and Autobiographical Sketches*. Cambridge: Cambridge University Press. First published in 1944.

Seddon, Philip J. 2017a. The Ecology of De-Extinction. *Functional Ecology* 31 (5): 992–995.

———. 2017b. De-Extinction and Barriers to the Application of New Conservation Tools. *The Hastings Center Report* 47 (4): S5–S8.

Shapiro, Beth. 2015. *How to Clone a Woolly Mammoth: The Science of De-Extinction*. Princeton: Princeton University Press.

Sheppard, Kathleen L. 2012. Between Spectacle and Science: Margaret Murray and the Tomb of the Two Brothers. *Science in Context* 25 (4): 525–549. https://doi.org/10.1017/S0269889712000221.

Shipani, Vanessa. 2019. Doudna's Confidence in CRISPR's Research Potential Burns Bright. *Quanta Magazine*, February 27. https://www.quanta-magazine.org/doudnas-confidence-in-crisprs-research-potential-burns-bright-20190227/. Accessed April 29, 2019.

Soulé, Michael. 1985. What Is Conservation Biology. *BioScience* 35 (11): 727–734.

Star, Susan Leigh, and James R. Griesemer. 1989. Institutional Ecology, 'Translations' and Boundary Objects: Amateurs and Professionals in Berkeley's Museum of Vertebrate Ecology, 1907–1939. *Social Studies of Science* 19 (3): 387–420.

Steffan, W., J. Grinevald, P. Crutzen, and J. McNeill. 2011. The Anthropocene: Conceptual and Historical Perspectives. *Philosophical Transactions of the Royal Society A: Mathematical, Physical and Engineering Sciences* 36 (1938). https://doi.org/10.1098/rsta.2010.0327.

Thacker, E. 2001. The Science Fiction of Technoscience: The Politics of Simulation and a Challenge for New Media Art. *Leonardo* 34 (2): 155–158.

———. 2003. Black Magic, Biotech and Dark Markets. In *Sarai Reader 03: Shaping Technologies*, ed. Sarai Collective. Delhi: Centre for the Study of Developing Societies (CSDS).

The Sydney Morning Herald. 2001. Tassie Tiger Pup That Holds the Hope for Revival. *The Sydney Morning Herald*, March 20. https://www.smh.com.au/national/tassie-tiger-pup-that-holds-the-hope-for-a-revival-20010320-gdfo0r.html. Accessed April 23, 2019.

Thomson, Helen. 2018. Hybrid White-Rhino Embryos Created in Last-Ditch Effort to Stop Extinction. *Nature*, July 4. https://doi.org/10.1038/d41586-018-05636-6.

Tumarkin, Nina. 1981. Religion, Bolshevism, and the Origins of the Lenin Cult. *Russian Review* 40 (1): 35–46.

Udall, S. 1963. *The Quiet Crisis*. New York: Holt, Rhinehart and Winston. https://archive.org/stream/quietcrisis00udal/quietcrisis00udal_djvu.txt. Accessed April 23, 2019.

Vankin, Deborah. 2016. Step Inside a Digital Storm: Andy Warhol's 'Rain Machine' Brought Back to Life After 45 Years. *Los Angeles Times*, October 20. http://www.latimes.com/entertainment/arts/la-et-cm-warhol-rain-machine-20161020-snap-story.html. Accessed September 23, 2018.

Venter, J. Craig. 2013. *Life at the Speed of Light: From the Double Helix to the Dawn of Digital Life*. New York: Viking Penguin Group.

Virslja, Z., S.G. Daniele, J. Silberies, F. Talpo, Y.M. Morozov, and A.M.M. Sousa. 2019. Restoration of Brain Circulation and Cellular Functions Hours Post-Mortem. *Nature* 568: 336–343. https://doi.org/10.1038/s41586-019-1099-1.

Watson, James D., and Francis H.C. Crick. 1953. Genetical Implications of the Structure of Deoxyribonucleic Acid. *Nature* 171: 964–967.

Whipple, Tom. 2014. Mammoth Challenge for Scientists Trying to Revive a Woolly Mammoth. *The Sunday Times*, March 22. https://www.thetimes.co.uk/article/mammoth-challenge-for-scientists-trying-to-revive-a-woolly-giant-l6crv2wsmz9. Accessed April 29, 2019.

Whiteley, A.R., S.W. Fitzpatrick, W.C. Funk, and D.A. Tallmon. 2015. Genetic Rescue to the Rescue. *Trends in Ecology & Evolution* 30 (1): 42–49.

Williams, Nigel. 2003. Death of Dolly Marks Cloning Milestone. *Current Biology* 13 (6): PR209–PR210.

Wilson, Clare. 2019. Pig Brains Have Been Partially Revived After Death—What Does This Mean. *The New Scientist*, April 17. https://www.newscientist.com/article/2200192-pig-brains-have-been-partly-revived-after-death-what-does-this-mean/. Accessed April 29, 2019.

Winter, Allison. (2013. Scientists Pursue 'Jurassic Park' Style Resurrections for Extinct Animals. *E&E News*. https://www.eenews.net/stories/1059977861/print. Accessed March 10, 2019.

Yirka, Bob. 2015. Researchers Take Another Step in Bringing Back a Woolly Mammoth. *Phys.org*, March 23. https://phys.org/news/2015-03-wooly-mammoth.html. Accessed April 28, 2019.

Yule, Jeffrey V. 2002. Cloning the Extinct: Restoration as Ecological Prostheses. *Common Ground* 1 (2): 6–9.

INDEX

© The Author(s) 2020
A. L. Fletcher, *De-Extinction and the Genomics Revolution*,
https://doi.org/10.1007/978-3-030-25789-7